THE GOD EFFECT

ALSO BY BRIAN CLEGG

Light Years: The Extraordinary Story of Mankind's Fascination with Light

The First Scientist: A Life of Roger Bacon

Brief History of Infinity: The Quest to Think the Unthinkable

THE GOD EFFECT

QUANTUM ENTANGLEMENT, SCIENCE'S STRANGEST PHENOMENON

BRIAN CLEGG

ST. MARTIN'S PRESS NEW YORK

 For information, address St. Martin's Press, 175 Fifth Avenue, New York, N.Y. 10010.

www.stmartins.com

Book design by Gretchen Achilles

Library of Congress Cataloging-in-Publication Data

Clegg, Brian.
The God effect : quantum entanglement, science's strangest phenomenon / Brian Clegg.—1st ed.
p. cm.
Includes bibliographical references and index.
ISBN-13: 978-0-312-34341-5
ISBN-10: 0-312-34341-8
1. Physics—Popular works. 2. Science—Popular works. I. Title.

QC24.5.C54 2006
530—dc22

2006042503

10 9 8 7 6 5 4 3 2

TO GILLIAN, REBECCA, AND CHELSEA

CONTENTS

PREFACE

If you thought science was a predictable, commonsense business—maybe even a little dull—you haven't encountered quantum entanglement. A physical phenomenon so strange and all pervasive that this book calls it "the God Effect," entanglement leaves common sense shattered. Inexplicable in normal terms, entanglement can reach instantaneously from one end of the universe to the other. Speculatively cited as the source of life and the mechanism of the mysterious Higgs boson, aka the God particle, entanglement has powerful potential uses from unbreakable encryption to teleportation. It's the strangest effect in all of science, yet hardly anyone has heard of it.

Once upon a time, science seemed straightforward. The British professor of natural history Thomas Huxley, whose support for evolution in the second part of the nineteenth century gained him the nickname "Darwin's Bulldog," described science as "nothing but trained and organized common sense." Yet the changes in science, and particularly physics, in the following century would prove him startlingly wrong.

Take quantum electrodynamics, the theory explaining the interaction of matter and light. This is how it was described in an open lecture by Richard Feynman, the brilliant twentieth-century American physicist whom many would consider one of the few true geniuses to flourish in the scientific community. (If you've never heard Feynman speak, imagine Tony Curtis reading these words):

The theory of quantum electrodynamics describes Nature as absurd from the point of view of common sense. And it agrees fully with experiment. So I hope you can accept Nature as She is—absurd.

I'm going to have fun telling you about this absurdity, because I find it delightful. Please don't turn yourself off because you can't believe Nature is so strange. Just hear me out, and I hope you'll be as delighted as I am when we're through.

The topic of this book, quantum entanglement, takes the absurdities and delights that inspired Feynman to a new level. Entanglement is remarkable in its own right, but even more amazing are the recent discoveries of real-world applications for this strange effect. Be prepared to encounter surprise and wonder.

GOING FURTHER

Quantum entanglement, is at the time of writing a field in furious development. There are new discoveries on an almost weekly basis. If you would like to read further, visit the *Popular Science* Web site, www.popularscience.co.uk, for features on the developing story of entanglement and recommendations of other books with which to explore the quantum world and the wider bounds of science and mathematics.

I have avoided reference numbers in the text to avoid breaking the flow, but the endnotes section, starting on page 247, gives detailed references for quotes, sources, and papers to help with further reading.

A NOTE ABOUT ALICE AND BOB

There is a long-standing convention among scientists dealing with quantum entanglement to refer to the owner of one end of an entangled process as Alice and the person at the other end as Bob. Usually Alice is trying to send Bob a message in this inane little drama.

The convention derives from cryptography, where the same names (along with other bit-part players, such as the eavesdropper, Eve) have been used in these roles for a good number of years. While the true origins of the names are lost in the mists of time, and they were obviously dreamed up to give more meaningful labels to a geometry diagram's dull point A and point B, it's hard for anyone of the right age to hear "Bob" and "Alice" without remembering the now very dated 1969 film *Bob & Carol & Ted & Alice* and wincing.

Where a convention has practical value it is worth maintaining, but Alice and Bob's role in entanglement is purely habitual and conveys nothing worthwhile—so to spare the pain of remembering the movie they will not appear again in this book.

THE GOD EFFECT

CHAPTER ONE

ENTANGLEMENT BEGINS

Laws are generally found to be nets of such a texture, as the little creep through, the great break through, and the middle-sized are alone entangled in.

—WILLIAM SHENSTONE, *Essays on Men, Manners, and Things*

Entanglement. It's a word that is ripe with implications. It brings to mind a kitten tied up in an unraveled ball of wool, or the complex personal relationship between two human beings. In physics, though, it refers to a very specific and strange concept, an idea so bizarre, so fundamental, and so far reaching that I have called it the God Effect. Once two particles become entangled, it doesn't matter where those particles are; they retain an immediate and powerful connection that can be harnessed to perform seemingly impossible tasks.

The word "quantum" needs a little demystifying to be used safely. It does nothing more than establish that we are dealing with "quanta," the tiny packets of energy and matter that are the building blocks of reality. A quantum is usually a very small speck of *something*, a uniform building block normally found in vast numbers, whether it's a photon of light, an atom of matter, or a subatomic particle like an electron.

Dealing in quanta implies that we are working with something that comes in measured packages, fixed amounts, rather than delivered as a continuously variable quantity. In effect, the difference between something that is quantized and something continuous is similar to the difference between digital information, based on quanta of 0s and 1s, and analog information that can take any value. In the physical world, a quantum is usually a very small unit, just as a quantum leap is a very small change—quite different from its implications in everyday speech.

The phenomenon at the heart of this book is a linkage between the incomprehensibly small particles that make up the world around us. At this quantum level, it is possible to link particles together so completely that the linked objects (photons, electrons, and atoms, for instance) become, to all intents and purposes, part of the same thing. Even if these entangled particles are then separated to opposite sides of the universe, they retain this strange connection. Make a change to one particle, and that change is instantly reflected in the other(s)—however far apart they may be. The God Effect has an unsettling omnipresence.

This unbounded linkage permits the remarkable applications of quantum entanglement that are being developed. It enables the distribution of a secret key for data encryption that is impossible to intercept. It plays a fundamental role in the operation of a quantum computer—a computer where each bit is an individual subatomic particle, capable of calculations that are beyond any conventional computer, even if the program ran for the whole lifetime of the universe. And entanglement makes it possible to transfer a particle, and potentially an object, from one place to another without passing through the space in between.

This counterintuitive ability of entanglement to provide an intimate link between two particles at a distance seems just as odd to physicists as it does to the rest of us. Albert Einstein, who was directly responsible for the origins of quantum theory that made entanglement inevitable, was never comfortable with the way entanglement acts at a distance, without anything connecting the entangled particles. He referred to the ability of quantum theory to ignore spatial separation as "*spükhafte Fernwirkungen,*" literally spooky or ghostly distant actions, in a letter written to fellow scientist Max Born:

> *I cannot make a case for my attitude in physics which you would consider reasonable . . . I cannot seriously believe in [quantum theory] because the theory cannot be reconciled with the idea that physics should represent a reality in time and space, free from spooky actions at a distance.*

Entanglement, as a word, seems to have entered the language of physics at the hand of scientist Erwin Schrödinger, in an article in the *Proceedings of the Cambridge Philosophical Society.* Interestingly, although German, Schrödinger was working and writing in English at the time—and this may have inspired his use of "entanglement"—the German word for the phenomenon, *Verschränkung,* has a rather different meaning than does his word choice in English.

The English term has subtly negative connotations. It gives a sense of being out of control and messed up. But the German word is more structured and neutral—it is about enfolding, crossing over in an orderly manner. A piece of string that is knotted and messed up is entangled, where a carefully woven tapestry has *Verschränkung.* In practice, neither word seems ideal. Quantum entanglement may lack the disorder implied by "entanglement," but it is much stronger and more fundamental than the pallid *Verschränkung* seems to suggest.

For Einstein, the prediction that entanglement should exist was a clear indicator of the lack of sense in quantum theory. The idea of entanglement was an anathema to Einstein, a challenge to his view on what "reality" truly consisted of. And this was all because entanglement seemed to defy the concept of locality.

Locality. It's the kind of principle that is so obvious we usually assume it without even being aware of it. If we want to act upon something that isn't directly connected to us—to give it a push, to pass a piece of information to it, or whatever—we need to get something from us to the object we wish to act upon. Often this "something" involves direct contact—I reach over and pick up my coffee cup to get it moving toward my mouth. But if we want to act on something at a distance without crossing the gap that separates us from that something, we need to send an intermediary from one place to the other.

Imagine that you are throwing stones at a can that's perched on a fence. If you want to knock the can off, you can't just look at it and make it jump into the air by some sort of mystical influence; you have to throw a stone at it. Your hand pushes the stone, the stone travels through the air and hits the can; as long as your aim is good (and the can isn't wedged in place), the can falls off and you smile smugly.

Similarly, if I want to speak to someone across the other side of a room, my vocal chords vibrate, pushing against the nearest air molecules. These send a train of sound waves through the air, rippling molecules across the gap, until finally those vibrations get to the other person's ear, start her eardrum vibrating, and result in my voice being heard. In the first case, the ball was the intermediary, in the second the sound wave, but in both cases something physically traveled from A to B. This need for travel—travel that takes time—is what locality is all about. It says that you can't act on a remote object without that intervention.

All the evidence is that we are programmed from birth to find the ability to influence objects at a distance unnatural. Research on babies has shown that they don't accept action at a distance, believing that there needs to be contact between two objects to allow one to act on the other.

This seems an extravagant assertion. After all, babies are hardly capable of telling us that this is what they think, and no one can remember how they saw the world in their first few months of life. The research technique that gets around this problem is delightfully cunning: babies are made bored by constant repetition of a particular scene, then after many repeats, some small aspect of the scene is changed. The babies are watched to see how they react. If the new movement involves action *with* visible contact, the babies get less worked up than if it appears to involve action at a distance. If a hand pushes a toy and it moves, the baby doesn't react; if a toy moves on its own, the baby does a double take. The inference that babies don't like the ability to act remotely is indirect, but the monitoring does appear to display babies' concern about action at a distance—the whole business feels unnatural.

Next time you are watching a magician at work, doing a trick where he manipulates an object at a distance, try to monitor your own reaction. As the magician's hand moves, so does the ball (or whatever the object he is controlling happens to be). Your mind rebels against the sight. You *know* that there has to be a trick. There has to be *something* linking the action of the hand and the movement of the object, whether directly—say, with a very thin wire—or indirectly, perhaps by a hidden person moving the object while watching the magician's hand. Your brain is entirely convinced that action at a distance is not real.

However, though action at a distance *looks* unreal, this doesn't rule out the possibility of its truly happening. We are used to having to overcome appearances, to take a step away from what looks natural, given extra knowledge. From an early age (unlike dogs and cats) we know that there aren't really little men behind the TV screen. Similarly, a modern child will have been taught about gravity, which itself gives the appearance of action at a distance. We know gravity works from a great range, yet there is no obvious linkage between the two bodies that are attracted to each other. Gravitation seems to offer a prime challenge to the concept of locality.

This idea of gravitational attraction emerged with the Newtonian view of the world, but even as far back as the ancient Greeks, before any idea of gravity existed, there was awareness of other apparent actions at a distance. Amber rubbed with a cloth attracts lightweight objects, such as fragments of paper, toward it. Lodestones, natural magnets, attract metal and spin around, when set on a cork to float on water, until they are pointing in a particular direction. In each case, the action has no obvious linkage to make it work. The attracted object moves toward the magnet—the floating lodestone spins and the static-charged amber summons its retinue of paper scraps as if by magic.

The Greeks had competing schools of thought on what might be happening. One group, the atomists, believed that everything was either atom or void—and, as nothing could act across a void, there had to be a continuous chain of atoms that linked cause to effect. Other Greek philosophers put action at a distance down to a sympathetic process—that some materials were inherently attracted to each other as one person attracts another. This was little more than a variant on the third possibility open to the Greek mind—supernatural intervention. In effect, this theory said there was something out there that provided an occult nudge to make things happen. This idea was widely respected in ancient times as the mechanism of the long-lasting if scientifically unsupportable concept of astrology, in which supernatural influence by the planets was thought to shape our lives.

Even though, nearly two thousand years later, Newton was able to exhibit pure genius in his description of *what* happened as a result of one apparent action at a distance—gravity—he was no better than the Greeks in explaining *how* one mass influenced another without anything connecting them. In his masterpiece, the *Principia Mathematica,* published in 1688, he said:

> *Hitherto, we have explained the phenomena of the heavens and of our sea by the power of gravity, but have not yet assigned the cause of this power. This is certain, that it must proceed from a cause that penetrates to the very centres of the sun and planets, without suffering the least diminution of its force; that operates not according to the quantity of the surfaces of the particles upon which it acts (as mechanical causes used to do) but according to the quantity of the solid matter which they contain, and propagates its virtue on all sides to immense distances . . .*
>
> *I have not been able to discover the cause of those properties of gravity from the phenomena, and I frame no hypothesis; for whatever is not deduced from the phenomena is to be called an hypothesis; and hypotheses, whether metaphysical or physical, whether of occult qualities or mechanical, have no place in experimental philosophy. In this philosophy particular propositions are inferred from the phenomena, and afterward rendered general by deduction . . . And to us it is enough that gravity does really exist, and acts according to the laws which we have explained, and abundantly serves to account for all the motions of the celestial bodies, and of our sea.*

This quote contains one of Newton's best-known lines, "I frame no hypothesis" ("*hypotheses non fingo*" in his original Latin). The modern translation of *Principia,* by Cohen and Whitman, points out that *fingo* was a derogatory term, implying making something up rather than the apparently neutral "frame." Newton was saying that gravity exists, but he wasn't going to provide a nonempirical guess at how it works. Some would continue to believe that gravity had some occult mechanism, on a par with astrology, but mostly the workings of gravity were swept under the carpet until Einstein came along.

One fundamental that came out of Einstein's work was that nothing could travel faster than light. We will revisit the reasoning behind this (and the implications of breaking Einstein's limit) in chapter 5. For the moment, though, relativity sounded the death knell for action at distance. It had been known since 1676 that light traveled at a finite speed, when the Danish astronomer Ole Roemer made the first effective determination of a velocity now set at around 186,000 miles per second. Einstein showed that action could not escape this constraint. Nothing, not even gravity, could travel faster than the speed of light. It was the ultimate limit.

We still don't know exactly how gravity works, but Einstein's limit was finally proved experimentally at the beginning of the twenty-first century—gravity *does* travel at the speed of light. If the sun suddenly vanished, just as we wouldn't see that it had disappeared for about eight minutes, we also wouldn't feel the catastrophic impact of the loss of its gravitational pull until then. Locality reigns.

Or at least that seemed to be the case, until experiments based on the work of an obscure physicist from Northern Ireland, John Bell, proved the existence of entanglement. Entanglement is genuine action at a distance, something that even now troubles many scientists. Of course, today we have a more sophisticated view of the universe—and have to face up to the fact that the concept of "distance" itself is perhaps not as clear and obvious as it once was. Theorist Berndt Müller, of Duke

University, has suggested that the quantum world has an extra unseen dimension through which apparently spatially separated objects can communicate as if they were side by side. Others imagine spatial separation to be invisible—in effect, nonexistent—to entangled particles. Even so, there is a powerful reluctance to allow that anything, however insubstantial and unable to carry information, could travel faster than light.

Although Einstein's objections to quantum theory based on its dependence on probability are frequently repeated (usually in one of several quotes about God not throwing dice), it was the breach of locality that really seemed to wound Einstein's sense of what was right. This is never more obvious than in a series of sharp handwritten remarks Einstein appended to the text draft of an article his friend Max Born had sent to him for comment:

> *The whole thing is rather sloppily thought out, and for this I must respectfully clip your ear . . . whatever we regard as existing (real) should somehow be localized in time and space . . . [otherwise] one has to assume that the physically real in [position] B suffers a sudden change as a result of a measurement in [position] A. My instinct for physics bristles at this. However, if one abandons the assumption that what exists in different parts of space has its own, independent, real existence then I simply cannot see what it is that physics is meant to describe.*

The phenomenon that challenges locality, that makes action a distance a possibility once more, the phenomenon of entanglement, emerges from quantum theory, the modern science of the very small. To reach the conception of entanglement, we need to trace quantum theory's development from a useful fudge to fix a puzzling phenomenon, to a wide-ranging structure that would undermine all of classical physics.

Max Planck, a scientist with roots firmly in the nineteenth century, started it all in an attempt to find a practical solution to an otherwise intractable problem. Planck, born in Kiel, Germany, in 1858, was almost put off physics by his professor at the University of Munich, Phillip von Jolly. Von Jolly held the downbeat view that physics was a dead-end career for a young man. According to von Jolly, pretty well everything that happened in the world, with a couple of minor exceptions, was perfectly explained by the physical theories of the day, and there was nothing left to do but polish up the results and add a few decimal places. Planck could have been tempted to build on his musical capabilities and become a concert pianist, but instead he stuck with physics.

It's a good thing he did. Von Jolly could not have been more wrong, and it was one of those "minor exceptions," the dramatically named ultraviolet catastrophe, that began the process of undermining almost all of von Jolly's "near perfect" physics, and that would elevate Planck to the pantheon of the greats. As far as the best calculations of the day could determine, a blackbody (a typical physicist's simplification: an object that is a perfect absorber and emitter of radiation) should emit radiation at every frequency, with more and more output in the higher frequencies of light, producing in total an infinite blast of energy. This clearly wasn't true. Objects at room temperature only gave off a bit of infrared, rather than glowing with an explosion of blue, ultraviolet, and higher-power radiation.

In 1900, Planck got around this seemingly impossible situation by dividing the possible emissions or absorptions of electromagnetic energy by physical matter into fixed units (quanta, as Einstein would soon call them). This was something that Max Planck would never truly be comfortable with. He wrote:

> *The whole procedure was an act of despair because a theoretical interpretation had to be found at any price, no matter how high that may be.*

For Planck, these "quanta" were not real. They were a vehicle to help him achieve a workable solution, a working method that had no direct connection to true physical entities. His quanta were, as far as he was concerned, an imaginary conceit. He considered them to be like numbers, when compared with physical objects. The number three (as opposed to "3," which is the symbol for the number three) isn't real. I can't show you three. I can't draw three, or weigh three. But I *can* show you three oranges—and the number proves very valuable when I want to make calculations concerning oranges. Similarly, Planck believed that quanta did not exist but made a valuable contribution to calculations on the energy of light and other forms of electromagnetic radiation.

There is an interesting parallel between Planck's attitude to quanta and the anonymous preface that was added to *De Revolutionibus*, the great work in which Nicholas Copernicus challenged the idea that the Sun traveled around the Earth. This tacked-on text, probably written by Andreas Osiander, the clergyman who supervised publication for the ailing Copernicus, is an introduction that dismisses the sun-centered theory of the book as a convenience for undertaking calculations that need bear no resemblance to reality. This was very similar to Planck's view of quanta.

Einstein, born twenty-one years later than Planck, was less fussy about detaching quanta from the real world. In a remarkable paper written in 1905 (the paper for which he later won his Nobel Prize), he suggested that light was actually made up of these quanta. Instead of its being continuous waves, he imagined it to be divided into minute packets of energy. Just how revolutionary Einstein's vision would be isn't clear from the title of the paper, *Über einen die Erzeugung und Verwandlung des Lichtes betreffenden heuristischen Gesichtspunkt* (On a Heuristic Viewpoint of the Creation and Transformation of Light). But revolutionary it was. Because, if there was one thing everyone was certain about—or at least everyone was certain about until Einstein changed everything—it was that light was a wave.

To be fair, Isaac Newton had always thought that light was made of particles, and his idea had kept going longer than it might otherwise because of the sheer momentum of the Newton name, but by the start of the twentieth century there was no contest between particles and waves. Not only did light exhibit behavior that made it a natural candidate for being a wave—bending around obstructions as the sea does around a breakwater, for instance—but Thomas Young had shown in a beautifully simple experiment in 1801 that light could produce interference patterns when passed through a pair of narrow slits. The mingled beams threw shadings of light and dark onto a screen, corresponding to the addition and subtraction of the ripples in the wave, just as waves did on the surface of water. No other explanation seemed capable of explaining light's behavior.

Figure 1.1. Young's experiment with two slits, showing light to be waves. The dotted lines show where waves are adding together to produce a bright line on the screen.

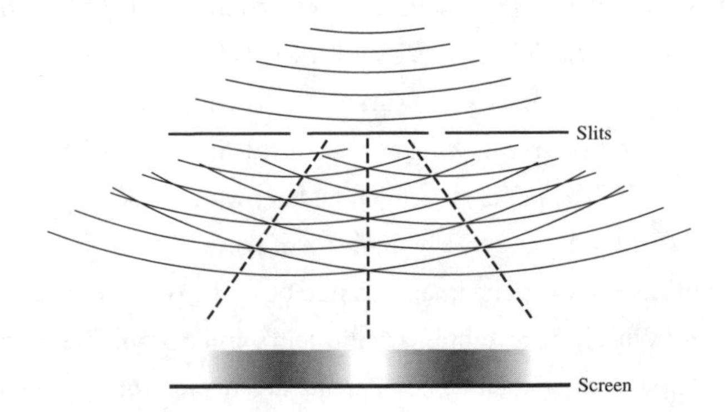

There was no way, for example, that the scientists of the time could imagine these interference patterns being developed by a series of particles. A particle had to follow a single path from source to screen. Passing a stream of particles through a pair of slits should result in two bright areas (one behind each slit) and large swathes of darkness, not the repeating dark and light patterns that everyone from Young onward could clearly see when they carried out the experiment.

In his paper on light, Einstein not only worked with Planck's quanta but showed that the radiation in a blackbody cavity behaved just like a gas of particles—he could apply the same statistical techniques that he had already successfully applied to gases. What's more, if light truly were made up of individual quanta rather than continuous waves, Einstein predicted it should be possible to generate a small electrical current when light was shone on certain metals, something that was suspected but had yet to be fully proved. This photoelectric effect really clinched the paper's significance.

Planck was no enthusiast for this promotion of his imaginary concept toward reality, and went so far as to criticize Einstein in a deeply condescending fashion. When Planck recommended the younger man for the Prussian Academy of Sciences in 1913, he asked that they wouldn't hold it against Einstein that he sometimes "missed the target in his speculations, as for example, in his theory of light quanta . . ."

The same year that Planck made this remark, Einstein's idea would be absorbed and amplified by the man who later became Einstein's chief sparring partner over quantum theory, and particularly over quantum entanglement, Niels Bohr.

Bohr, born in Copenhagen, Denmark, in 1885 (and so around ten years younger than Einstein), came from a family with deeply academic roots. His father was a professor of physiology and his younger brother became a math professor. After earning a doctorate in Copenhagen, Bohr had traveled to England, first to Cambridge and then to work in the northern industrial city of Manchester with the New Zealand–born physicist Ernest Rutherford, who had come to fame by discovering the atomic nucleus.

In 1913, Bohr devised a model of the atom's structure that relied on Einstein's quanta to explain its workings. His idea was to consider the atom, with its tiny but heavy central nucleus surrounded by much smaller electrons, as if it were a sun with its attendant planets in orbit. Although this model of the atom would be discarded relatively quickly in the scientific world, it became very popular with the general public, particularly in the 1950s, when symbols of a nucleus with electrons whirling around it proliferated endlessly—and even now, children are often taught this as their initial view of the atom, probably a mistake, as it is very difficult then to shake off this image and we now know that atoms just aren't built that way.

Of course, technically this inaccuracy is true of any description we make of physical phenomena, particularly those that take place on scales that are too large or too small to easily comprehend. Our explanations of the workings of the world, from the atom to the big bang, are all "models," the scientific equivalent of a metaphor. Metaphors (and models) can be misleading if taken too literally, as a conversation in the animated movie *Shrek* demonstrates.

The hero, Shrek, makes the comment that ogres are like onions because they have layers. His friend, Donkey, takes the metaphor too literally and assumes that ogres are like onions because they stink, or make you cry, or grow little hairs when left out in the sun. Metaphors and models can be dangerous if you take them too literally, or extend them too far—and sometimes they can be more misleading than any value gained from such an illustration.

Science writer John Gribbin once firmly criticized physicist Nick Herbert for saying he felt dishonest "whenever I draw for schoolchildren the popular planetary picture of the atom; it was known to be a lie even in their grandparents' day." Gribbin responded sternly, "Is it a lie? No! No more so, at least, than any other model of atomic reality." But this is unfair on Herbert—the fact is that some models are better than others.

The planetary model that is being criticized was better than the older "plum pudding" model that imagined negatively charged electrons to be scattered through a homogenous mass of positive charge, like fruit in a plum pudding. But equally, the planetary model is more misleading than newer alternatives that don't pretend that electrons behave in the neatly ordered manner of planets in stately motion around a sun.

In fact, as soon as Bohr came up with his planetary model, there was a problem (and this is where Einstein's quanta came into the picture). A satellite in orbit—the Earth around the Sun; the Moon or an artificial satellite traveling around the Earth—is constantly accelerating. This doesn't mean it gets quicker and quicker, because this is a different kind of acceleration. Ever since Newton, we've known that a moving body will travel in a straight line at a constant speed unless you apply a force to it. The satellite wants to fly off in a straight line, out of orbit. It is only the constantly applied force of gravity that pulls it out of the straight line and around the curve. And a body that has a force applied to it is said to accelerate.

In this case, the result is not straight-line, linear acceleration like a drag racer accelerating down the track, but angular acceleration. With this type of acceleration, the speed remains the same but the direction changes. Velocity, the true measure of rate of movement, comprises both speed and direction. The velocity is changing because, though the speed remains the same, the direction is constantly being modified. That works fine with a satellite and, in a stable orbit, if there isn't any resistance, it could keep going around forever. But if this really were the case with an electron, there would be a different problem that would doom it to spiral inward and crash into the nucleus.

Bohr knew that an electron normally pumps out light when it is accelerated. That inevitably means losing energy—all electromagnetic radiation, like light, carries a certain amount of energy. An electron that is orbiting, that is accelerating, should emit a stream of light, rapidly losing energy, before crashing destructively into the nucleus. This doesn't happen. (Thankfully, or every atom of matter in existence would have self-destructed within a tiny fraction of a second of being created.) So Bohr had the clever idea of putting his electrons on imaginary tracks.

Instead of being able to swing around the nucleus in any old orbit, Bohr imagined that electrons were constrained to travel on fixed circuits, still confusingly called orbits. Once on a track—in what Bohr called a stationary state—the normal rules did not apply: it was as if the imaginary track held in the photons and stopped energy from leaking out. The electrons could jump from one orbit to another—giving out or absorbing a quantum of light—but could not live anywhere in between. It wasn't possible for electrons to gradually drift down and crash into the nucleus; they could only make instantaneous leaps between fixed orbits. These jumps between different tracks, gaining or losing a quantum of energy with each jump, were called quantum leaps. Bohr had taken the atom digital.

Niels Bohr will come back into the story, but first his ideas had to be transformed and upgraded by the young Turks of quantum theory—Prince Louis de Broglie, Werner Heisenberg, Erwin Schrödinger, and

Paul Dirac. De Broglie inverted Einstein's idea that light—usually seen as a wave—could be thought of as particles, by showing that elementary particles like electrons could behave as if they were waves. Heisenberg abandoned Bohr's visually appealing orbits to produce matrix mechanics, a totally abstract mathematical description of the processes involved. Schrödinger came up with an alternative view, a description of the way de Broglie's waves changed with time, known as wave mechanics—and Dirac showed that Heisenberg and Schrödinger's approaches were not just consistent but totally equivalent, pulling the two together as quantum mechanics.

All was not rosy in the quantum mechanical garden, however. If Schrödinger's wave equations were taken as literal descriptions of the behavior of quantum particles (something he hoped for, as he hated the abstract nature of Heisenberg's matrices, with no accompanying picture of what was happening), there was a problem. If a particle like an electron were literally a wave, following the behavior specified by Schrödinger's equations, it would have to spread out in all directions, rapidly become ridiculously huge. And there were other complications in the way Schrödinger's equations used imaginary numbers and needed more than three dimensions when more than one particle was involved. The solution to making these wave equations usable came from another of the new generation of physicists, Einstein's friend Max Born.

Born may have been as close as anyone ever was to Einstein socially, but he brought into quantum theory the apparently simple concept that would cause Einstein and others so much trouble—probability. To make Schrödinger's wave equations sensibly map onto the observed world, he suggested that they did not describe how an electron (for instance) moves, or the nature of an electron as an entity, but rather provided a description of the *probability* that an electron would be in a particular place. The equations weren't a distinct picture of an electron but a fuzzy map of its likely locations. It was as if he had moved our image of the world from an accurate modern atlas to a medieval muddle with areas labeled "here be electrons."

It is from this introduction of probability into the quantum world that Heisenberg's uncertainty principle would emerge. Werner Heisenberg showed that quantum particles had pairs of properties that were impossible to measure simultaneously in absolute detail (properties are just aspects of an object that can be measured like its mass, position, velocity and so on). The more accurately you knew one of the properties, the less accurately you could measure the other. For example, the more closely a particle's momentum was known, the less accurately its position could be determined. (Momentum is the mass of the particle multiplied by its velocity [directional speed], something physicist John Polkinghorne describes in a matter-of-fact way as "what it is doing.") At the extreme, if you knew exactly what momentum a quantum particle had, it literally could be positioned anywhere in the universe.

A good way of picturing the uncertainty principle is provided by Doctor Peet Morris, of Oxford University's Computing Laboratory. Imagine you take a photograph of an object that is flying past at high speed. If you take the picture with a very quick shutter speed, it freezes the object in space. You get a good, clear image of what the object looks like. But you can't tell anything from the picture about its movement. It could be stationary; it could be hurtling past. If, on the other hand, you take a photograph with a slow shutter speed, the object will show up on the camera as an elongated blur. This won't tell you a lot about what the object looks like—it's too smudged—but will give a clear indication of its movement. The trade-off between momentum and position is a little like this.

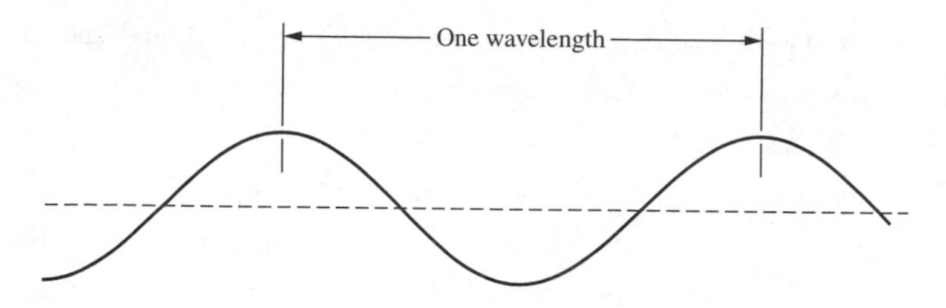

Figure 1.2. A wavelength in the progress of a beam of light.

This uncertainty principle seems to become obvious when you think about trying to make this measurement for real. Say you used a light beam to accurately measure the position of an electron. One of the definitive properties of light is its wavelength—the distance in which the imagined motion of a light wave goes through a complete ripple and gets back to the equivalent point in its travel.

The wavelength of the light determines how accurately it can be used to measure position. But the shorter the wavelength, the more energy the light carries—and the more effect being hit by that light will have on the momentum of the electron. The mere act of looking at a quantum particle changes it—something that would become a central tenet of quantum theory.

This description of the uncertainty principle as a side effect of measurement comes from Heisenberg's paper on the topic that used the example of a microscope in which the light that was being used to observe a particle disrupted it. However, this interpretation isn't as straightforward as it seems. In Heisenberg's original example, the act of measurement causes the uncertainty—so by implication, if no measurement were made, the momentum and position could have absolute values.

This seems to have been Heisenberg's original understanding, when he presented his microscope example to the great Niels Bohr. Heisenberg is said to have ended up in tears as Bohr pointed out that, though the uncertainty principle is correct, the microscope example is hopelessly misleading. It assumes an underlying reality—it pictures the electron as traveling along a clear, specific path until the light disrupts it. But as far as quantum theory is concerned, it isn't like that. Born had shown that Schrödinger's wave equation described not what the particle itself did, but the *probability* of its taking a particular route. An electron does not follow a specific path—all that can be said is that, when a measurement is taken, certain values are obtained. These give a guide (subject to Heisenberg's principle) of where the electron might be at a particular moment, but do not imply that it followed a specific, known path to get there.

Although this verges on the sort of philosophical conundrum that inclined seventeenth-century philosophers to ask whether a tree falling in a forest made any sound if there was no one there to hear it, it is important to the whole debate on what is really happening at the quantum level, and would be part of the reason behind the split that divided Einstein from many of his colleagues.

In fact, Einstein had been uncomfortable about the role of randomness in the physics of the very small from an early point in his career. Austrian quantum physicist Anton Zeilinger has pointed out that it was as early as 1909, in a presentation at a Salzburg gathering of scientists and medics (the magnificently named Gesellschaft der Deutschen Naturforscher und Ärtze Salzburg), that Einstein commented that he was discomforted (*unbehagen*) by the role that random events played in the new physics.

This was very obvious from a series of letters between Einstein and Max Born. On April 29, 1924, he wrote:

> *I find the idea quite intolerable that an electron exposed to radiation should choose of its own free will, not only its moment to*

jump off, but also its direction. In that case, I would rather be a cobbler, or even an employee in a gaming house, than a physicist.

Einstein could not accept this randomness: he felt there should be a strict, causal process underlying what was observed. As far as he was concerned, the electron jumped out of the metal it was in at a time and in a direction that could have been predicted, had all the facts been available. Quantum theory disagreed, saying it wasn't ever possible to know when the electron would pop out, or in what direction. Similarly, quantum theory assumed that a particle *didn't have* a position until the measurement was made—it was the act of measurement that transformed its position from a probability to an actual value. And the same went for the other properties of the particle.

By December 4, 1926, Einstein was sufficiently irritated by the topic to write his famous words:

Quantum mechanics is certainly imposing. But an inner voice tells me that it is not yet the real thing. The theory says a lot, but does not really bring us any closer to the secret of the "old one." I, at any rate, am convinced that He is not playing at dice.

The problem was this matter of probability and statistics. There was nothing new about apparently random effects in physics, and Einstein himself had made significant use of statistics and probability in his work; for example, in describing the effect called Brownian motion, whereby pollen and other small particles jump about in a fluid, cannoned into by fast-moving molecules. But his assumption had always been that there were real values underlying those probabilities.

For instance, probability theory shows that the chance of throwing a coin and getting a head is 50:50—there is a 1 in 2 probability. But any particular coin throw will have a real, specific outcome. It will produce a head or a tail, and when the probability is worked out over many coin throws, each of those throws would have a specific outcome. Now Born and Bohr were saying that, in quantum physics at least, the reality had to be thrown away. *All* that existed was the probability. Einstein could not accept this—he made a huge distinction between what was actually happening and the tools that could be used to predict the outcome. To see what his problem was, it's worth indulging in a quick detour into the nature of probability.

"Statistics" and "probability" are terms that are often used with more enthusiasm than accuracy. The Victorian British prime minister Benjamin Disraeli said, "There are three kinds of lies: lies, damned lies, and statistics."

This contempt for statistics dates back to the original use of the word, when it was a political statement of facts about a country or community (the "stat" part is as in "state"). The origin of the complaint is the way statistics can be used to support almost any political argument—but political distaste should not be allowed to conceal the value that the statistical method has for science. Statistics gives us an overview of a large body of items that we couldn't possibly hope to monitor individually—almost any measurement on a gas in the real world (pressure, for instance) is statistical, because it combines the effects of all the many billions of gas molecules present.

Probability, on the other hand, is about chance. It usually describes something that may or may not happen. In any particular case, there is usually a single actual outcome, but we can give that outcome a probability. So when the weather forecast tells us there's a 50 percent probability of rain, in practice either it will rain or it won't. We just know that the chances of it happening are pretty evenly balanced. Let's look at probability and statistics applying to the specific example we've already seen—tossing a coin.

Imagine you have a coin, and toss it a hundred times in a row. Each time you toss the coin, you note down the outcome. Probability tells us that each time you throw the coin you've a 50:50 chance of getting a head or a tail. So probability predicts that, on average, after a hundred throws, you will get fifty heads and fifty tails. If we actually count heads and tails, you might end up with forty-eight heads and fifty-two tails. These statistics tell us what happened. Probability is the likeliness of what will happen in the future—statistics describe the actual outcome. Probability tells us which combinations are more or less likely, while statistics open up reality. If we took more and more tosses, the statistical outcome would lead us toward deducing the 50:50 probability. The two are connected, but are only the same for certain if we had an infinitely large statistical sample, something that doesn't exist in the real world.

Einstein thought that the microscopic world should be like this, too. It's rather as if we had a clever machine that tossed the coins and automatically sent all the heads to one closed hopper and all the tails to another. We could then view the results by weight. The closest we would get to describing what has happened is a probability deduced from the statistics—we could see from the weights that there were forty-eight heads and fifty-two tails, and after many experiments could deduce the 50:50 probability. We never get to see a coin, or a toss, but we would know that the coins did exist in the machine. Similarly, Einstein was convinced that somewhere underneath the probabilities lay reality. He believed that, behind the screen of probability, there was a set of actual values that it was impossible to see. Whether these values, sometimes called hidden variables, existed would trigger the whole quantum entanglement debate.

Although Einstein expressed his concern to Born, the originator of the probabilistic interpretation of Schrödinger's wave equations, it was with the older Niels Bohr that he engaged in the great battle over the acceptability of quantum theory. It didn't matter to Bohr that the more recent versions of the theory had pushed his atomic orbits out of the way; he was a great supporter of quantum mechanics—in fact, as a result of the support it got from Bohr's Danish center of operations, the most widely accepted understanding of quantum phenomena is still often refered to as the "Copenhagen interpretation."

Einstein developed the habit of teasing Bohr by turning up at conferences with thought experiments that challenged the validity of quantum theory. He was particularly successful at winding up his Danish counterpart at the Solvay Congresses, large-scale scientific gatherings that brought together many of the big names of physics at that time.

Einstein's first shot across Bohr's bow came at the fifth Solvay Congress in Brussels, in 1927. After hearing the big names talk about the latest developments in quantum theory, Einstein described a simple experiment that he thought highlighted a fundamental problem with the theory. He imagined firing a beam of electrons at a narrow slit. On the other side of the slit, the electrons (behaving like waves) were diffracted, spreading out from the slit rather than simply passing through in a straight line. Einstein then imagined a semicircular piece of film that was marked by the impact of the electrons.

According to quantum theory, it isn't possible to say where the individual electrons would be until they hit the film. Schrödinger's equations described the probability of an electron's being in any particular spot, but it is only at the moment of impact that one spot on the film

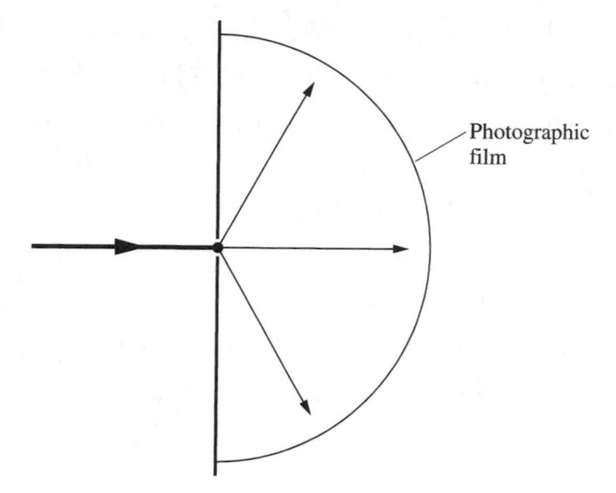

Figure 1.3. Einstein's electron beam thought experiment.

would blacken and a measurement take place. Einstein wasn't happy with this. He tried to imagine the moment an electron hit the film. The way he envisaged the experiment, if quantum theory was true, it was as if every point on the film had a random chance of turning black, defined by the probability distribution. Anywhere was in danger of suddenly turning black. But the instant one bit of film did actually register an impact, somehow all the other bits of film had to immediately know not to turn black. It seemed to him as if an instantaneous communication had to connect all parts of the semicircle, telling each bit of the film whether or not to respond.

This time around, Bohr was not particularly bothered by Einstein's remarks—in fact, he found the idea totally confusing. He commented, "I feel myself in a very difficult position because I don't understand what precisely is the point which Einstein wants to [make]. No doubt it is my fault."

Einstein didn't give up but continued with two rather more complex thought experiments, producing them as he sat at breakfast in the hotel before the conference began. The details of these thought experiments don't really matter—what Einstein believed he had done was to show that, with some clever manipulation (for example, using a shutter to only allow a short pulse of electrons or photons through a slit, and by combining what was known about the shutter and the slit with the information gained by measuring the spread of the particles), he could defy the uncertainty principle, knowing more about the location and momentum of the particles than he was allowed to. If the uncertainty principle could be shown to be faulty, then quantum theory was in serious trouble.

This time, Bohr took the threat seriously—but Einstein's new ideas didn't take much opposing and Bohr was ready with a solution by dinner on each occasion. The problem was, Einstein hadn't taken into account the uncertainties that were present in measuring just what the shutter and slit were doing. He had assumed that these could be known absolutely. If, however, they, too, were subjected to the uncertainty principle, this extra information was lost. The thought experiment was entirely consistent with the uncertainty principle and couldn't be used to disprove it.

It was another three years before the two great minds had another chance to come head-to-head on the subject, at the next Solvay Congress, in 1930, held as usual in Brussels. The topic of the conference was magnetism, but this didn't stop Einstein from presenting one of his breakfast-table challenges to quantum mechanics. And this time he felt that he was onto a winner. Einstein had come up with a very clever thought experiment that appeared to challenge the reality of the uncertainty principle, and hence quantum theory.

Einstein's experiment (it should be emphasized that this wasn't a real experiment anyone intended to carry out, just a mental vehicle for testing the theory) consisted of a box with a source of radiation inside. On the wall of the box, a shutter covered a hole. The shutter was opened for a very brief time, during which a single photon shot out of the hole.

It was known by then that, though a photon of light didn't have any mass (if you could imagine stopping the photon and weighing it [impossible in practice, as Einstein had already established light has to always travel at light speed]), its energy when moving produced an effective mass that could be derived from $E=mc^2$. So Einstein envisaged weighing the box before and after the photon was produced. This would give him a value for the energy of the photon that he could make as accurate as he liked. But he could also measure the time when the shutter was opened, to great accuracy. And this combination would allow him to know time and energy to better accuracy than he should know with the uncertainty principle, which applies to time and energy in just the same way as it does to position and momentum.

Bohr could not see the flaw in this devilish contraption. An observer at the time described Einstein as walking quietly away from the meeting with a "somewhat ironical smile" while Bohr trotted excitedly beside him.

Next morning, however, Bohr was ready to come back to Einstein with a counterargument, ironically using Einstein's own general relativity against him. He imagined a particular example of Einstein's apparatus (though the same argument can be applied to different arrangements with a bit of work). The box with the shutter is hung from a spring. This forms the weighing machine that will detect the change in mass. A clock inside the box opens and closes the shutter.

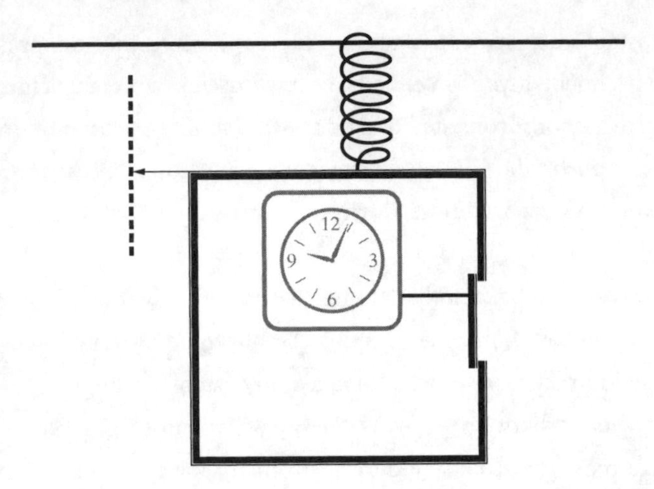

Figure 1.4. Bohr's version of Einstein's photon experiment.

Now as the photon shoots out the box, the whole apparatus will move upward, indicating the change in weight and hence the change in energy. There is uncertainty in this value *and* uncertainty in the time measured, as clocks in motion in gravity run slowly, according to Einstein's general theory of relativity. The two uncertainties combine to produce exactly the desired results of the uncertainty principle. Once again, Einstein had to accept defeat. However he tried to attack the outcomes of quantum theory, the ability of the measurements to interfere with each other managed to destroy his challenge. Yet, still he believed that it *should* be possible to dig deeper and uncover the hidden reality.

Bohr was never quite sure whether these attacks from Einstein were purely technical or driven by an urge to irritate him (though there is no evidence that Einstein was motivated by anything other than a disquiet with quantum theory, he was a very accomplished teaser). Even years later, in 1948, Bohr was clearly unnerved by Einstein's challenges. Physicist Abraham Pais recounts how he was attempting to help Bohr put together some information on his disputes with Einstein: At the time Bohr, was visiting the Institute for Advanced Study at Princeton and was using the office adjacent to Einstein's. (Technically, the office he was in *was* the one allocated to Einstein, but Einstein preferred the more cramped confines of the room that should have belonged to his assistant.)

Bohr was supposed to be dictating his text to Pais but, as often happened, he was having trouble stringing together a sentence. The concepts were there in his head, but Bohr found it difficult to formulate the wording. The eminent scientist was pacing around the table in the middle of the room at high speed, almost running, repeating, "Einstein . . . Einstein . . ." to himself. After a little while, he walked to the window and gazed out, repeating now and then, "Einstein . . . Einstein . . ." At that moment, the door opened very softly and in tiptoed Einstein, himself. He signed to Pais to keep quiet with what Pais later described as "his urchin smile" on his face.

It appears that Einstein had been ordered by his doctor not to buy any tobacco. Treating this injunction as literally as he could, Einstein decided that he couldn't go to the tobacconist but that it would be okay to raid Bohr's tobacco, which was in a pot on Bohr's table. After all, by stealing the tobacco, he was sticking to the injunction not to buy any. As Einstein crept into the room, Bohr was still facing the window, still occasionally muttering, "Einstein . . . Einstein . . ."

On tiptoe, Einstein made his way toward the desk. At that point Bohr uttered a final, firm "Einstein!" and spun round to find himself face-to-face with his longtime opponent, as if the incantation had magically summoned him. As Pais commented:

> *It is an understatement to say that for a moment Bohr was speechless. I myself, who had seen it coming, had distinctly felt uncanny for a moment, so I could well understand Bohr's own reaction.*

But back in 1930, after his successful dismissal of Einstein's challenge, Bohr was feeling rather more comfortable in the security of quantum theory. It would be five years before Einstein would strike back again in a way that would throw Bohr into total confusion, though ironically it would make it possible to prove that the strange quantum world really did exist.

Over those five years, Einstein put together the elements of a thought experiment that would enable him to separate the two measurements linked by the uncertainty principle so firmly that he believed one would not be able to influence the other. That way, he became increasingly convinced, he could cheat uncertainty and shake quantum theory to its foundations. He first tried out the concept verbally on Léon Rosenfeld at the 1933 Solvay Congress, after hearing Bohr's latest thoughts on the quantum theory, but this time Einstein did not intend to launch a lightweight challenge over breakfast. This was to be a formal, scientific paper.

The move away from teasing Bohr by casually dropping nightmare problems in his lap reflected the increasingly dark situation in Europe. Hitler's Germany drove Einstein reluctantly to the United States, where he set up residence in what would be his home for the rest of his life, Princeton, New Jersey. The Institute for Advanced Study, recently founded in 1930, seemed the ideal location. Set up by Louis and Caroline Bamberger, it brought together experts in the theoretical sciences (there were, and still are, no laboratories), math, and history in a relaxed environment where there was an opportunity to work without the distraction of students and lectures. It provided all the good parts of a university (from the academic's viewpoint) without the time-wasting chores.

It was from here, with two collaborators, Boris Podolsky and Nathan Rosen, that Einstein would produce the paper that brought the implications of entanglement out into the open. Published in *Physical Review* on

May 15, 1935, and called "Can Quantum-Mechanical Description of Physical Reality Be Considered Complete," the paper became universally known by the initials of its authors, EPR. Its intention was to smash quantum theory by showing just how incredible the resultant entanglement would be. Einstein was out to destroy the opposition.

CHAPTER TWO

QUANTUM ARMAGEDDON

I wave the quantum o' the sin,
The hazard of concealing;
But, och! it hardens a' within,
And petrifies the feeling!

—ROBERT BURNS,
"Epistle to a Young Friend"

EPR. Three initials that any physicist would immediately identify with Einstein, Podolsky, and Rosen's paper on quantum entanglement. Einstein had been harrying leading quantum scientists, and particularly Niels Bohr, for years. Now in 1935 with his paper—and despite the three names at the top of EPR, no one doubted that it embodied Einstein's opinion—he had declared open warfare. The English wording of the paper might not have been perfect (Einstein soon distanced himself from the overcomplex text) but the entertaining skirmishing of the Solvay Congresses had been pushed aside with a written declaration of war.

Over the years this one paper, which effectively spawned the whole concept of entanglement, has been misused and misinterpreted in pretty well every way possible. It has been said that Einstein and his colleagues were proved wrong by Bohr's response to the paper. This is not true. Others have claimed that the paper wasn't real science at all, but more philosophical posturing. Also not true. Yet others claim that the logical inevitability of EPR proves that Einstein was right all along—yet subsequent experiment would show that this viewpoint was also incorrect.

We need to look in a little more detail at this document that has produced many echoes throughout the subsequent history of physics, but let's start with the demolition of those false assertions. What the paper does is to present the reader with a quandary, and the quandary is very real. To this extent, EPR is absolutely correct and Einstein wasn't wrong. EPR points out that either quantum theory has gaps or locality doesn't work. Either there are really hidden pieces of information in the universe that quantum theory insists are fuzzy uncertainties, or, alternatively, locality—the idea that two things separated at a distance can't influence each other without something passing between them—is untrue.

However this doesn't amount to a walkover for the Einstein supporters. It only takes a couple of minutes' reading of the paper to realize that its authors believed that the solution to the quandary was the first of the options: that gaps existed in the theory. This is implied by the title of the paper, "Can Quantum-Mechanical Description of Physical Reality Be Considered Complete?," and is made obvious in the text. So, in an ironic reflection of quantum superposition, the feature of the quantum world that allows particles to be in more than one state at a time, the EPR paper managed to be both correct and incorrect at the same time. It is right that either quantum theory has a hole in it, *or* the idea of locality is flawed, but it is wrong in suggesting that the first of these is the only sensible option.

Toward the end of the paper, the authors do briefly bring up the possibility that two separated systems could have a direct influence on each other—that locality could be breached—only to dismiss this idea in a very summary fashion: "No reasonable definition of reality could be expected to permit this," says EPR. Unfortunately, as we now know, the workings of the world at the quantum level are anything but reasonable. "Reasonable" doesn't enter into the equation. Which means it isn't enough to appeal to common sense, as Einstein et al. were doing. Such an assertion needs experimental proof.

Einstein's doubts about the wording of the paper were, frankly, well substantiated. The writing style doesn't do justice to his argument. Einstein's English was not brilliant at the time—he had only been in the United States for two years—and it appears that Boris Podolsky was the actual author. Podolsky's words lack the directness and careful avoidance of unnecessary complexity that Einstein favored. The basic argument of EPR is obscured by an unnecessarily messy structure and a loose use of terms. There is reasonable evidence that Einstein never saw the paper before it was published, merely discussing the ideas with Podolsky and Rosen, which Podolsky then wrote up.

Cutting through the complexity, though, the underlying proposition is simple enough. Einstein and his colleagues imagined a particle breaking down into two others, a common enough occurrence in quantum physics. The two new particles shoot off in opposite directions, each with an equal and opposite momentum just as Newton had foretold should happen. The initial particle wasn't moving, so the momentums of the two new particles had to cancel out, because you can't produce momentum out of nowhere. What's interesting about this setup is that each particle can tell us something about the other. Measure the distance one has traveled and you know how far the other has gone. Measure one particle's momentum and you know the momentum of the other.

According to quantum theory, the properties of a particle aren't fixed until we measure them. Remember it's not just that a property has a secret value all the time, and we now discover it by making the measurement—quantum theory depends on the counterintuitive concept that there are only probabilities of various values until we make a measurement, at which point a particular, actual value (give or take a degree of uncertainty) is established.

Say we measure the momentum of the first particle. Because of the neat symmetry of the experiment, we immediately also know the momentum of the second particle. But according to quantum theory, neither particle had a fixed momentum until the moment the first particle was measured. Now, immediately, we know the value for both particles, however far apart they have traveled.

At this stage comes the clincher. At the instant we measured the first particle's momentum, how did the second particle "know" what momentum it should have? If its momentum was, until that moment, just a range of probabilities rather than a particular fixed value, what caused it to jump to a particular actual momentum—the same momentum as the first particle, but in the opposite direction? In this challenge, Einstein brings in the concept of locality that we met in the previous chapter. It would seem that it is only by instant action at a distance that one particle could influence the other. After all, we could wait as long as we like before making the measurement, so the two particles could be light-years apart. Assuming (as Einstein did) that it's impossible for any communication between the two instantaneously, the only deduction we can make is that the second particle already had that momentum.

That was actually as much as was required to make the assertion that either there's a gap in quantum theory or the whole idea of local reality collapses. The most obvious unnecessary complexity in EPR is that it then goes on to talk about making a second, separate measurement of the position of the particles.

To be precise, EPR suggests doing a second, similar experiment, but this time measuring where one of the particles is at a particular time. Again, we can find out as exactly as we like (within the limits of our equipment) how far the particle has traveled, assuming we ignore momentum, and from this can also deduce how far the second particle has gone. As soon as we measure the position of the first particle, we instantly know the position of the second one, which implies that its location already had that real value, unless there is some mystical, instant, long-distance communication between the two. So, concludes EPR, the second particle already had a fixed momentum *and* position, something that directly contravenes quantum theory. As the paper puts it:

> *If, without in any way disturbing a system, we can predict with certainty (i.e. with probability equal to unity) the value of a physical quantity, then there exists an element of physical reality corresponding to this physical quantity.*

EPR set out to prove that it was possible to predict values of momentum and position of a distant particle exactly, though not both at the same time—unlike Einstein's earlier attempts, EPR wasn't trying to disprove the uncertainty principle. If the prediction were possible, and hence there was something real underlying these measurements, quantum theory had a problem.

Yet we have to remember that what they actually managed is to show that either there is something missing from quantum theory, or that local reality is an incorrect assumption. EPR was riskily prepared to dismiss action at a distance with that remark, "No reasonable definition of reality could be expected to permit this." We will see in chapter 5 just why this was such a strong requirement for any sensible theory as far as Einstein was concerned.

In practice, the example used (probably selected by Rosen) was not the best because the underlying math happens to be more complicated when dealing with momentum and distance than it would be with other properties of particles. And as we have seen, there was really no need to involve two properties. All that was needed was to show that a measurement of one property, which immediately fixed the value for a second particle, could be made when the particles were too far apart for the information to travel from one to the other before the outcome could be checked. Another U.S. physicist, David Bohm, later introduced a variant of the EPR thought experiment in which, instead of momentum and position, the spin of a similarly produced pair of particles is measured, making it a simpler and cleaner demonstration of the quantum theory quandary.

The spin of a quantum particle is another of those "an ogre is like an onion" models owing more to metaphor than reality. One of the reasons the model of electrons orbiting the nucleus as planets orbit the sun was a handy one, is that electrons have a property that resembles angular momentum. Just as ordinary momentum describes the tendency that a body moving in a straight line has to keep on going, so angular momentum, or more properly orbital angular momentum, is the "oomph" with which one body orbits another. The more it has, the harder it is to stop it. And the particles had yet another varying property, so rather like assuming ogres, like onions, go brown in the sun and grow little white hairs, it was arbitrarily decided (to fit with the model) that this other property should be called spin angular momentum, or "spin" for short.

In the normal world, spin momentum is the momentum of a body spinning on its own axis. Carried away by the picture of something that moved like the earth going around the sun, the theorists imagined this new property that particles could have was due to the electron spinning on its axis. It isn't. They could just as easily have called the property saltiness or bounce or frangibility, but they decided to call it spin.

At its most basic, spin is a way of distinguishing particles that comes in two flavors, spin up and spin down. (Up and down describes the direction of the axis. Imagine a right hand cupping the electron with the thumb pointing up. If the electron was imagined to be spinning in the direction of the fingers, counterclockwise as seen from the top, the spin is up, the direction of the thumb.) But remember, this does not reflect the actual behavior of anything. In practice, all we can say is that particles with different spins will shift in different directions when put through a detector that's a like a tunnel with a north magnetic pole one side and a south pole on the other.

In case, despite all the warnings, you are in danger of accepting that an electron really is spinning around a particular axis, the reality of taking measurements should shatter your illusions. Whatever direction you measure the spin of the electron in it will either be "up" or "down" in that direction—it can't be somewhere in between. With a real spinning object like the Earth, if we tried to measure its spin on something other than a line through the poles, we would say it had something between spin up and spin down, a factored combination of the two. But spin is entirely different with a quantum particle. It can only be up or down, whatever direction you measure it in. Spin is a quantized, a digital property. The probability of a measurement coming up with "up" or "down" will vary, but there are only two possible outcomes.

Just as the original EPR paper described linked momentum, so the spin of the second particle is linked to that of the first when two particles are created entangled. Either the value of the spin is mysteriously communicated to the second particle when a measurement is made on the first, or it had a value all along. The main advantage in the spin variation of EPR is that the math is much simpler, but the basic concept is the same.

Léon Rosenfeld, a young colleague of Bohr's, was in Copenhagen when the EPR paper was published, and according to Abraham Pais's account of the reaction in the Bohr camp, Rosenfeld was baffled the next morning when Bohr burst into his office, smiling profusely, and shouting with every appearance of the onset of insanity: "Podolsky, Opodolsky, Iopodolsky, Siopodolsky, Asiopodolsky, Basipodolsky." Even Bohr's explanation that the babbling was a parody of a line spoken by a servant in Holberg's play *Ulysses von Ithaca* failed to make it seem any more reasonable.

The version that Pais tells of Bohr's moment of shocked realization contrasts a little with Rosenfeld's own account, which has him informing Bohr of the EPR paper rather than the other way around, and describes how Rosenfeld witnessed Bohr's rapidly growing concern:

> *As soon as Bohr heard my report of Einstein's argument, everything else was abandoned: we had to clear up such a misunderstanding at once. We should reply by taking up the same example and showing the right way to speak about it. In great excitement, Bohr immediately started dictating to me the outline of such a reply. Very soon, however, he became hesitant. "No, this won't do, we must try all over again . . . we must make it quite clear . . ." So it went on for a while, with growing wonder at the unexpected subtlety of the argument. Now and then he would turn to me: "What* can *they mean? Do* you *understand it?"*

After some weeks of floor pacing, muttering, and discomfort, Bohr was able to respond to EPR on a number of levels. One was simply to say that the approach went against the principle of complementarity, a clumsily named idea at the heart of Bohr's interpretation of quantum theory. This principle maintains that where there are two mutually exclusive ways of observing a phenomenon, either is entirely viable until you take one approach—at that point the other becomes impossible. So, for instance, complementarity says that light can be considered to act like a particle or a wave, but as soon as you do an experiment that requires it to behave as one of these, it can't be treated as the other.

Similarly, complementarity says that as soon as you do an experiment to measure the momentum of a particle, you can't simultaneously measure the distance traveled. Bohr was unhappy that EPR's discussion of momentum and distance traveled went contrary to complementarity. But this argument of Bohr's misses the point of EPR (perhaps he can't be blamed, given its clumsy wording). The combination of measuring momentum and distance traveled was a red herring. Even in the original EPR formulation, the aspect that was supposed to be strange was the way the momentum of the second particle is fixed instantly the first is observed, however far apart they are. Or the way one distance was fixed by the other. But not both together.

Initially there seems to be another opportunity for EPR to shake uncertainty. Why not measure the momentum of the *first* particle and the distance traveled by the *second*? In each case we are only measuring a single property, so there would seem to be no problem in finding out both the momentum of one particle and the position of the other with absolute accuracy—and then combining the two to disprove Heisenberg's uncertainty principle. In fact, the challenge is an illusion. It's true we might know the momentum of the second particle (say) just before its position was measured, but as soon as we make the measurement of distance, anything we knew about the momentum ceases to be valid: fundamentally to quantum theory, the very act of measurement makes a change.

In truth, Bohr did not come up with a satisfactory response to Einstein. This was almost inevitable, as EPR wasn't really *wrong.* As we have seen, its basic premise that either there's something missing from quantum theory or local reality is breached by entangled particles is true. It's also likely that Bohr was incapable of pulling EPR to pieces, because fundamentally he agreed with one of the key assertions of the EPR authors. Along with Einstein, he believed that nonlocal action was impossible.

This wasn't something Bohr went out of his way to make clear. He commented in his response to EPR that taking a measurement on one particle would influence "the conditions which define the possible types of prediction" for the other. But this seems to have been part of an attempt to smother EPR in a blanket of confusion. After all, the same year he had written that to support nonlocality would mean that "we would truly find ourselves in irrational territory." Bohr believed that it was "completely incomprehensible" that an action at one place in an experiment should influence what happened in a totally separate place. For all his bluster, Niels Bohr was at heart a closet EPR supporter.

Perhaps because of this, the response that Bohr made to EPR was more political spin than a practical scientific challenge. He threw up a cloud of confusion, helped along by EPR's complex phrasing, which led many to dismiss it, or to regard it as more a matter of semantics than physics.

By now, the positions that Bohr and Einstein took lacked sufficient common ground to allow them to communicate sensibly. Physicist David Peat made a telling observation by comparing their stances with a clash between Cézanne and a realist painter, both engaged in producing a still life of a pile of oranges on a tablecloth. Each of the artists could argue that the other's view lacked something. Each was portraying a particular view of reality, one based on the static image, the other on a more holistic view of the nature of the oranges and their setting. Neither was truly "wrong," yet neither was complete. The same went for Bohr and Einstein.

It was the EPR paper that really brought the concept of entanglement to life. As we saw in the first chapter, Erwin Schrödinger wrote a piece for the *Proceedings of the Cambridge Philosophical Society*, inspired by Einstein, Podolsky, and Rosen's work. In it he stated:

> *When two systems, of which we know the states by their respective representatives, enter into temporary physical interaction due to known forces between them, and when after a time of mutual influence the systems separate again, then they can no longer be described in the same way as before, viz. by endowing each of them with a representative of its own. I would not call that* one *but rather* the *characteristic trait of quantum mechanics, the one that enforces its entire departure from classical lines of thought. By the interaction the two representatives [the quantum states] have become entangled.*

There was the word *entangled*, used for the first time to describe this (in Schrödinger's words) characteristic trait of quantum mechanics. It is lucky Schrödinger did use this term, as the same paper sees him referring to the phenomenon in a much less reader-friendly way, perhaps because of the tendency in his mother tongue of German to lump words together in a versatile but clumsy mix. Schrödinger points out that entanglement doesn't just link two specific measurements but a whole infinity of possible measurements. This, he comments, is with "no mnemotechnical help whatsoever"—so we should all be thankful that we are able to talk about entanglement and not an amnemotechnical link.

Schrödinger was almost as uncomfortable with the mysterious remote connection that entanglement implied for two entangled particles as was Einstein. To get around any suggestion that this fracture in locality could break the bounds of relativity, Schrödinger said that he believed the entanglement process would only ever be very short-lived—in fact he thought that entanglement could only exist over distances where the time taken for light to travel from one point to the other was negligible compared with the time taken for other changes in the system to take place. This speculation has since been shot down by practical experiment—and there was never any theoretical basis for the idea, other than Schrödinger's concern to preserve the spirit as well as the concrete reality of relativity.

EPR would be Einstein's last shot at unbalancing Bohr and his precious quantum theory. From here on we leave Einstein behind as we follow his unwanted creation on its way to deliver remarkable things in the form of quantum entanglement. But before passing Einstein by, it is worth remembering that his opposition to quantum theory did not go away just because he stopped coming up with new challenges. On September 7, 1944, nearly ten years after the EPR paper, he wrote to Max Born:

> *I [believe] in complete law and order in a world which objectively exists, and which I, in a wildly speculative way, am trying to capture. I firmly* believe, *but I hope that someone will discover a more realistic way, or rather a more tangible basis than it has been my lot to find. Even the great initial success of the quantum theory does not make me believe in the fundamental dice game . . .*

And as late as 1952, just three years before his death, Einstein was still scathing on the subject of quantum theory:

> *This theory reminds me a little of the system of delusions of an exceedingly intelligent paranoic, concocted of incoherent elements of thoughts.*

For Einstein, quantum theory (and by implication entanglement) would never make sense.

The enigma contained in Einstein, Podolsky, and Rosen's paper became widely known as the "EPR paradox." Nathan Rosen was never happy with this description. At a conference in the 1980s to mark the fiftieth anniversary of EPR he announced, "This term is unjustified, there is no paradox." But Rosen was being unnecessarily defensive. Einstein himself called the problem that the EPR paper raised a paradox in a letter to Erwin Schrödinger. (He also complained that EPR hid the real problem when, as we have seen, the challenge could be made as easily with one property of the particles as the two used in EPR. He commented to Schrödinger that dealing with the two properties of momentum and position "*ist mir wurst*," literally meaning "is sausage to me"—idiomatic German for "I couldn't care less about it.")

It has been suggested by some commentators that the use of "paradox" makes it sound as if EPR showed up an illogicality or absurdity in quantum theory—but this confuses a paradox with a fallacy. The point of a paradox is that it is a real result, an actual occurrence that challenges common sense. A paradox makes the eyebrows raise, but it is technically correct. I would suggest that EPR became regarded as a paradox because it appeared to challenge quantum theory, yet in practice seemed not to achieve anything more than raised eyebrows. And it was in this paradoxical state of "interesting but irrelevant to real science" that entanglement was to remain until a theoretical breakthrough brought it into the real world.

This doesn't mean that EPR was the last challenge to the version of quantum theory championed by Bohr, the "Copenhagen [or standard] interpretation." David Bohm, the man behind the spin variant on EPR, devised an ingenious approach, based on an idea of de Broglie's, called the quantum potential method that seemed to explain both normal classical effects and quantum phenomena without giving up the concept of real values for the likes of position and momentum. Bohm's theory isn't without problems—or it would have displaced the mind-twisting Copenhagen interpretation—but it is a valid challenge.

Equally, there have been several variants of the "many worlds" theory, where each measurement splits off new universes (one for each possible outcome of the quantum state), or the measurement switches "our universe" between a set of already existing parallels. Each interpretation has its own relatively small but enthusiastic following. But though entanglement is an important part of quantum theory, it isn't necessary to settle on a particular *interpretation* of the theory to follow the development of entanglement. In the end, these are just explanatory models—if any of the theories denies entanglement it runs into serious problems with the experimental evidence. Evidence that followed the inspirational breakthrough of John Bell, an obscure physicist from Northern Ireland.

Bell is probably the most important scientist no one but his colleagues has ever heard of. Unlike movie stars or politicians, whose very existence depends on reaching the public eye, scientists tend to get on with their jobs in the shadows, however much their work influences our lives. A few names have entered public consciousness, of course—Isaac Newton and Albert Einstein; Charles Darwin and Richard Feynman. If *Star Trek: The Next Generation* is anything to go by, we also should feature Stephen Hawking in our list (Hawking appears as a hologram in a poker game where the *Enterprise*'s android, Data, takes on the greatest scientists of history). But the vast majority of scientific workers, however able, remain in the shadows of obscurity. With them sits John Bell, the bearded, red-haired theoretical physicist who in 1964 became pivotal in establishing the reality of quantum entanglement.

Bell was born thirty-six years earlier on July 28, 1928, in Belfast, Northern Ireland. In this city of religious divides, the Bell family was Protestant from the moderate Church of Ireland (part of the same Anglican communion as the Episcopal Church in the United States), but the Bells showed little of the prejudice that has ruined many lives in the north of that divided country—the family had friends from all parts of the religious and political spectrum.

The Bells had a family history that was more practical than academic. It must have been something of a surprise when the eleven-year-old John junior, then called Stewart (his middle name, used by Bell until he went to university), proudly announced that he wanted to be a scientist. But Bell's mother, Annie, encouraged Stewart to carry on with his schooling well beyond fourteen, the age at which his brothers and sister quit the schoolroom to look for work. His further education was first at Belfast Technical High School and then Queen's University, Belfast. In Annie's words, she wanted him to have a life "where he could wear his Sunday suit all week!"

Bell had obvious talent and enthusiasm for learning. He was nicknamed "the Prof" at home because of his tendency to share his scholarship with anyone who would stand still long enough. But he was all too aware of the burden his education had placed on his family, and rather than look for an academic post after graduation—poorly paid in the 1950s—he took a job with the UK's Atomic Energy Research Establishment, based at Harwell in the rural heart of England. It was there that he met his Scottish wife-to-be, Mary Ross, recruited to the same team.

By 1960, Bell's enthusiasm for the Harwell work had waned. An opportunity arose for the Bells to take posts at CERN, which they snapped up. CERN (Conseil Européen pour la Recherche Nucléaire) is a vast international research establishment working on high energy particles, located nominally in Geneva but in fact straggling over (or, rather, under) the border between Switzerland and France. Best known now for the spin-off success of its electronic communication vehicle the World Wide Web and its role in Dan Brown's thriller *Angels and Demons,* CERN is a place where the basic components of the universe are battered together with immense energy in an attempt to analyze their makeup and understand their characteristics.

The work at CERN was in particle physics, but Bell had always had a spare-time interest in the fundamentals of quantum theory, and he took the opportunity of a year's sabbatical that found him touring Stanford, Wisconsin, and Brandeis universities in the United States, to follow that interest into some original theory. By the end of 1963, he was ready to produce a paper that would lead to today's remarkable applications of entanglement.

It wasn't immediately obvious when Bell published his paper in an obscure (and soon to collapse) journal called *Physics* that it was anything more than an abstract piece of theory that would interest very few readers. Bell's paper showed how to make an indirect measurement that should prove whether or not one of the predictions of quantum theory was true. Bell had derived a measurement that demonstrated whether two entangled particles really could influence each other at any distance, or whether quantum theory had gaping holes in it. It took five years for anyone to notice that Bell's work opened the way to an experimental exploration of this strange feature of the quantum world, and at least another ten more before the outcome was firmly established beyond reasonable doubt.

Although he was certainly fascinated by it, Bell himself had mixed feelings about quantum theory. He once commented, "I hesitated to think it might be wrong, but I *knew* that it was rotten." By this he seems to have meant that whatever lay at the heart of quantum theory wasn't described very well—the explanations of quantum phenomena simply didn't make sense. This attitude is perhaps clearest in his comment to fellow physicist Nick Herbert that, in producing his paper, he was delighted "in a region of wooliness and obscurity to have come across something hard and clear."

It was not the existence of quantum theory per se that offended Bell, but the fuzziness of what was said about it. He also felt that there was something missing. His natural inclination was to side with Einstein against the vast majority of physicists, in his criticism of the ideas at the heart of the quantum world. The purpose of his intervention in 1964 was to devise a new thought experiment that made it clearer that only if quantum theory was wrong could you have local reality—"reality" here meaning that there were true, if hidden, values of what was being measured rather than a fuzzy probability distribution. Bell once commented, "I felt that Einstein's intellectual superiority over Bohr, in this instance, was enormous; a vast gulf between the man who saw clearly what was needed, and the obscurantist." According to physicist Andrew Whitaker, Bell considered Bohr's response to the EPR paradox incoherent.

What is perhaps strange is that this view of Bell's seems to have been an unusual one at the time. Practically everyone writing between the 1960s and the 1990s on EPR and Bell takes the line that the whole world assumed the outcome of EPR was that Einstein was wrong and Bohr was right. In fact, this assertion, still being made by some writers today, is extremely presumptuous. The majority of the world would not have heard of EPR or Bohr, and would assume that Einstein was right, because . . . well, because he was Einstein, so he had to be right.

However, many in the scientific community, happy that quantum theory explained the observed results and not wanting to be drawn into philosophical arguments, were prepared to accept Bohr's point of view, without thinking about what this implied. As long as the result matched experiment, you could sweep under the carpet any discomfort caused by the interpretation.

Bell's apparent contempt for Bohr has to be taken in context. I don't believe any scientist would deny that Niels Bohr was a great physicist. It's highly unlikely that John Bell would have suggested any different. Bohr made a huge contribution to our understanding of the physical world. It's just that most onlookers would rather he had not resorted to his obscuring tactics and had stuck to more practical things. No one ever accused Bohr of being a great communicator.

Bell's thought experiment that eventually revived interest in entanglement was based on an extension of the spin measuring variant of EPR. He envisaged a particle breaking into two parts that shoot off, as usual, in opposite directions. At a good distance from each other, waiting for the two particles, are measuring devices that will be used to discover the spin of each of the particles. But the two measuring devices are not lined up the same way. One might measure whether spin was up or down at 22 degrees from the vertical, the other might be 54 degrees away from that, or at any other angle.

Now, if the two detectors were lined up the same way, Bell showed that it's possible to set up "hidden variables" to explain what is happening. And he could get hidden variables to work if the result for one particle depends on the alignments of both detectors (something that locality would seem to exclude as there wouldn't be time for a particle at one end of the experiment to find out the alignment of the detector at the other end). However, he proved it was impossible to come up with a variable or set of variables that would cope with totally independent directions settings for the detectors with no means of a particle "knowing" about both alignments.

Bell's theorem, as this paper's results became known, shows that you can't produce the predictions of quantum theory without dropping locality from the picture. Not just that, he described a situation where the outcome would be different depending on whether local reality applied or quantum theory's predictions were true. If you compared results for the two detectors over a wide range of angles, quantum theory predicted one result; local reality, with its hidden variables, predicted another. With the right experimental setup, all you would have to do was measure a set of values. If they fell statistically outside a certain range (referred to as Bell's inequality), then Bell's theorem was true and locality went out of the window.

John Bell had dared take on some of the greatest minds of the twentieth century. Like all the truly original thinkers of science, Bell was prepared to ignore anyone else's viewpoint when he thought he was right. His idea was not to prove Einstein wrong—nothing would have pleased Bell more than to have shown that Einstein was right and quantum theory was flawed—but with admirable scientific objectivity, rather than state any personal preference, he had laid out clear alternatives, only one of which could hold.

Most important of all, Bell's described an experiment that, while still technically a thought experiment, a mental challenge rather than a practical laboratory setup, was one that could in principle be carried out in the real world. This is in total contrast to Einstein's flamboyant challenges to quantum theory, which were never intended to be practical. Bell had moved the battleground over entanglement from the mind to the laboratory. And the outcome would be even fiercer disputes.

CHAPTER THREE

TWINS OF LIGHT

O'er the rugged mountain's brow
Clara threw the twins she nursed,
And remarked 'I wonder now
Which will reach the bottom first?'

—**HARRY GRAHAM**, *Ruthless Rhymes for Heartless Homes*

John Bell had provided a mechanism for putting EPR to the test, an experimental basis for choosing between quantum theory and local reality, but writing this paper was still more of a hobby than real work to him—and anyway, Bell was a theorist rather than an experimental scientist, with neither the opportunity nor the inclination to follow up his paper in the laboratory. Turning his ideas into a real-life experiment with widely accepted results that would make or break quantum theory and prove whether entanglement truly provided a spooky connection at a distance would be the work of a maverick young French scientist, Alain Aspect.

Aspect is anything but the stereotypical unworldly scientist with pocket protector, thick glasses, and a profound lack of knowledge of anything outside his lab. Born in 1947 in the southwest of France near the great wine-growing region of Bordeaux, Aspect was drawn to Paris to study physics. A big man with an impressive, flowing mustache, you might guess that he was a French TV presenter or even a barman, but not a scientist. His independent nature took him, after his doctorate, out to Cameroon, where for three years he undertook backbreaking physical effort in the African sun as an aid worker.

Most of this central African country, stretching a narrow footprint from the Gulf of Guinea in the west up to Lake Chad in the north, was briefly a French colony, and it was the remnants of its French associations that brought the young Aspect there in 1971. In the hot evenings, exhausted from his day of manual labor, Aspect would let his mind wander over what he regarded as the most interesting challenges that were facing physics, and in particular the controversies surrounding quantum mechanics.

It was a subject that had interested him in college, but he had never been totally comfortable with the way it was presented. Now, given a chance to think things through his own way, he was able to assemble a mental structure that matched his ideas, to put together a personal view of the quantum world uninfluenced by the fashion of the day.

It might seem odd to mention fashion in association with physics. It's easy to assume that science is purely objective and stands nobly above trends and fads. In science, surely, something is either right or wrong, not in or out of fashion? Well, no. The fact is that science is just as much subject to fashion as are hemlines—except instead of being driven by the arbitrary whims of a few influential designers, the current fashion in science is decided by a combination of which fields are currently considered to be high opportunity (the ones most likely to advance an academic career quickly), and by the political desires of those responsible for funding research.

In the early seventies, exploring the foundations of quantum theory was as far out of fashion as knickerbockers. The hot stuff in the hard sciences all seemed to involve smashing particles together with greater and greater energy, or developing new theories of cosmology, dramatic (and sometimes unlikely) speculation on how the universe was formed. Particle physics was particularly attractive because it produced completely new results. I was an undergraduate at the time, and it seemed almost every week one of our lecturers would excitedly announce the discovery of a new particle. Even better, particle physics involved building enormous, shiny machines that looked good on TV. The politicians could see what they were getting for their (or rather your) money.

Quantum theory, on the other hand, had the feeling of a field where there was little left to do, where theory matched experiment with boring predictability. The quantum world might still seem strange and new to us but, to physicists of the time, it was an old man's game. Even the term "mechanics" in "quantum mechanics" made it seem antiquated. Sure, the interpretation wasn't clear, but that didn't make any difference to practical outcomes. Quantum mechanics was the sort of thing that the previous generation had got all excited about. It was about as fashionable as your dad's taste in pants.

In his African backwater, Alain Aspect could not have been pulled along by the latest hot topic even had he wanted to. As it happens, his very personal approach was one that didn't encourage following the pack. Instead, he was intrigued by the Einstein, Podolsky, and Rosen paper, by then well over thirty years old. Somehow he had come across not only the original paper, which at least was a well-known relic of physics history, but also John Bell's intriguing but obscure extension of the EPR concept in the direction of practical testing. Here was a challenge thrown down to Aspect that, for the moment at least, he could only meet in his head.

It may well be that having that time to think things through, rather than rushing into the lab, was to prove Aspect's greatest weapon in meeting the challenge of entanglement. By the time he returned to Paris, he was determined to settle the outcome of Bell's inequality once and for all.

Aspect was not the first to attempt to turn Bell's ideas into a practical experiment. In 1969, four American physicists published a paper that took Bell's original idea further down the road to reality. This paper was a relatively rare example of parallel development leading to a positive outcome. Often in the history of ideas, two individuals or groups come up with a similar concept at around the same time. Once the components of an idea are present in the air it only takes a spark of creativity to set them off, and quite often that spark occurs in more than one place.

At their worst, parallel developments can lead to insults, acrimony, and lawsuits. When Isaac Newton and German mathematician Gottfried Wilhelm von Leibniz both developed calculus at the same time, Newton was convinced it was a matter of plagiarism and orchestrated a campaign to have Leibniz declared a cheat. Leibniz, a fellow of London's prestigious Royal Society, complained that he was being maligned, and in response the society set up an eleven-man committee to decide who had priority. The committee's report, written by the society's president, came down strongly in favor of Newton, something Leibniz never accepted. His reluctance might have been colored by the fact that the president of the Royal Society at the time was Isaac Newton.

Such disputes weren't limited to the mathematical world. When Thomas Edison in the United States and Joseph Swan in the Great Britain both demonstrated an electric lightbulb based on a carbon filament at around the same time, there was no question of information leaking between the two, it was a pure simultaneous development. For the hard-bitten Edison, though, this was nothing less than a commercial declaration of war. Even though Swan had first devised the bulb eight months earlier than Edison, he had not taken out a patent, so Edison brought in the lawyers.

Often in such circumstances, financial muscle can triumph over what's right, but the courts accepted Swan's priority in this particular discovery. Not only was Edison unable to sue Swan, he was forced to set up a joint company—the Edison and Swan United Electric Light Company—to exploit the invention. Such bitter rivalries are equally common in the boardroom and the university common room. Scientists, too, can be very jealous of their ideas and wary of sharing any possible glory.

Given these circumstances, it's easy to imagine the chagrin of the Boston-based team of Abner Shimony and his doctoral student, Mike Horne. They had been hard at work on a paper describing a means to test the outcome of Bell's theorem, only to discover that another scientist had just given a talk on the exact same topic. Colleagues suggested that the best approach for Shimony and Horne to take was to pretend they'd never heard of the other work—neither of them had attended the conference where the test was described—and to go ahead and publish their paper anyway. If they had gone down that path, they may well have ended up in a fierce battle for priority. But instead, they decided to call their rival.

The man behind the conference speech was John Clauser from Columbia University. Clauser had been one of the earliest to spot the significance of Bell's remarkable paper, and had even suggested testing it out for real when he was a graduate student at the California Institute of Technology. The reaction of his professor, the great Richard

Feynman, to this suggestion was apparently to throw him out of his office. Feynman was a man of amazing insights, but even he had his off days.

For Shimony and Horne, the idea of calling Clauser wasn't without risk—many other scientists would have been inclined to say, "Tough, but I went public first." Clauser, though, was happy to work with the others, doubly so as Shimony had already gotten Richard Holt, a graduate student at Harvard, on board. Holt was an experimenter who was capable of putting the theorists' ideas into action. The united four would make two significant steps forward. They eliminated an unwarranted assumption from Bell's original work and they changed their experimental vehicle from electrons to more easily handled photons of light.

Although photons do have spin, the easiest directional property to deal with is a spin-related phenomenon called polarization. This is a familiar concept from Polaroid sunglasses. Filters like the lenses of these glasses only let through light that is polarized in a particular direction. This is doubly useful. Not only does a polarizing filter cut down the level of light passing through (as randomly generated light is polarized in every direction imaginable, so a fair amount of the light is blocked) but also light reflected off a road surface tends to be polarized in some directions more than others, so by arranging the Polaroid glasses to weed out such photons, the glasses reduce reflections for drivers.

Polarization is usually visualized by thinking of light traveling along as a wave, like a ripple sent down a rope. Tie one end of rope to a ring and hold the other. You can move the rope up and down, sending vertical ripples along it, move the rope side to side, producing horizontal ripples, or shake it in any direction in between. If these ripples were instead light rays, the direction the light ripples in perpendicular to its direction of travel is its polarization. This is fairly obviously another ogres-and-onions thing—even when thought of as a wave, light is not a single wave but interacting electrical and magnetic waves at right angles to each other—and, when it's seen as a photon, it is even less obvious what polarization is. But the fact remains that polarization is a property of the photon that has a known direction associated with it, at right angles to the direction of travel.

When a beam of photons passes through a polarizer at an angle, it gives a graphic demonstration of just how strange the quantum world is. Let's say that we start off with a beam of photons in a natural, randomly polarized state, which we send through a polarizer set at 45 degrees, halfway between the vertical and the horizontal. Then the beam hits a vertical polarizer. What happens next?

If the light had been an ordinary wave, we might imagine that after passing through the first polarizer, the beam contained only photons polarized at 45 degrees—not horizontal or vertical. But when a photon hits the vertical polarizer, some pass through; some are absorbed and stop dead. We can say statistically that half of the photons that got through a 45-degree polarizer will get through the vertical polarizer, but it's impossible to say *which* photons. It's not that after going through the 45-degree polarizer, half the photons are vertical and half are horizontal. They are all in a "superposed" state of both horizontal and vertical, with a 50:50 chance of becoming one or the other when they hit the vertical polarizer.

If you imagine the polarizer as a slot through which a coinlike photon has to pass, just how remarkable this is becomes apparent. Quantum theory says that the photon is in *both* states (horizontally and vertically polarized) at once with an equal probability of selecting either one when it hits the polarizer. Only then do half the photons get

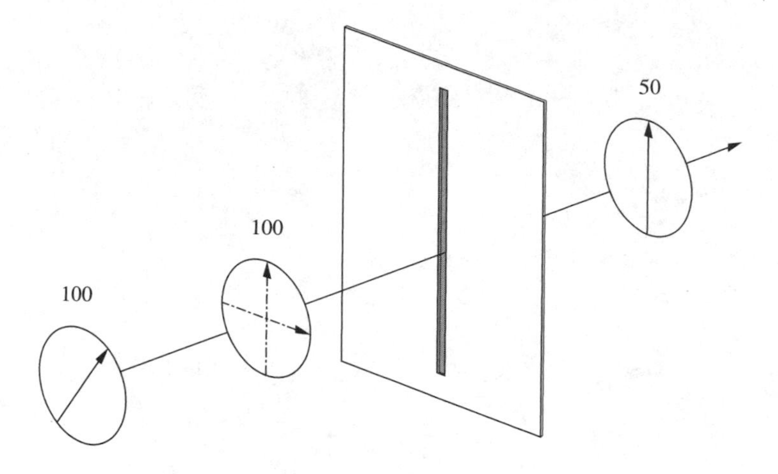

Figure 3.1. A photon polarized at 45 degrees combines horizontal and vertical with 50 percent probability.

through (now polarized vertically) and half fail. There's no way of knowing beforehand whether a particular photon will or won't pass through.

For example, if we send one hundred photons through at 45 degrees, that's equivalent to one hundred photons each with a 50:50 chance of becoming horizontal or vertical when it hits the polarizer, so on average fifty horizontally polarized photons will emerge.

It's important to emphasize this because it's so important (and so strange). Though at first glance it might appear that all that is happening here is that half the photons *always were* polarized horizontally and half always were polarized vertically (so photons polarized at 45 degrees are a mix of the two), a simple experiment that can be done with three lenses from Polaroid sunglasses will show that this isn't the case.

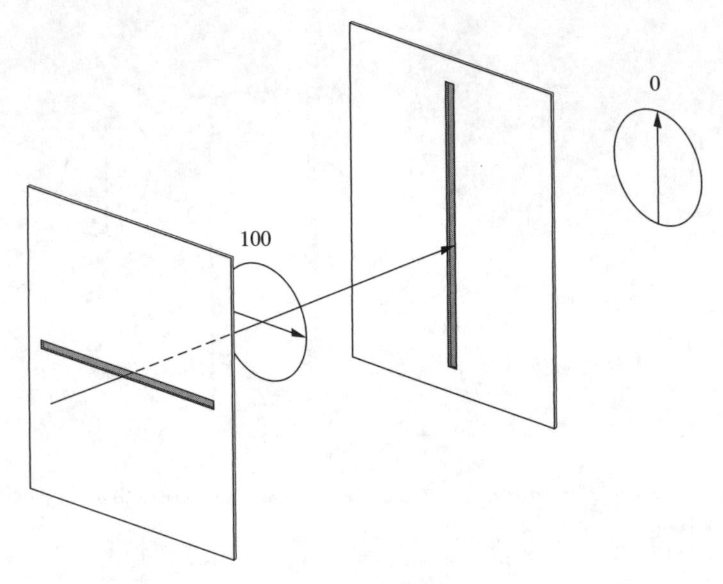

Figure 3.2. Opposing polarizers stopping all photons.

If we send light first through one polarizer, then through another at 90 degrees to it, nothing gets through. In the example above, all 100 percent of the photons are horizontally polarized by the first polarizer, so nothing gets through the second, vertical polarizer.

With two polarizers, nothing came out the far side. But take a third polarizer, turn it to a 45-degree angle, and place it in the middle, and something very strange happens. Once more you can see through your Polaroid sunglasses (if considerably more dimly). Photons of light are getting through. If the diagonal polarizer worked by letting through a mix of vertically and horizontally polarized photons, then after passing through this polarizer all the photons that started out horizontally polarized would still be horizontally polarized. Instead, that central polarizer produces photons in both states at once, with a 50:50 proba-

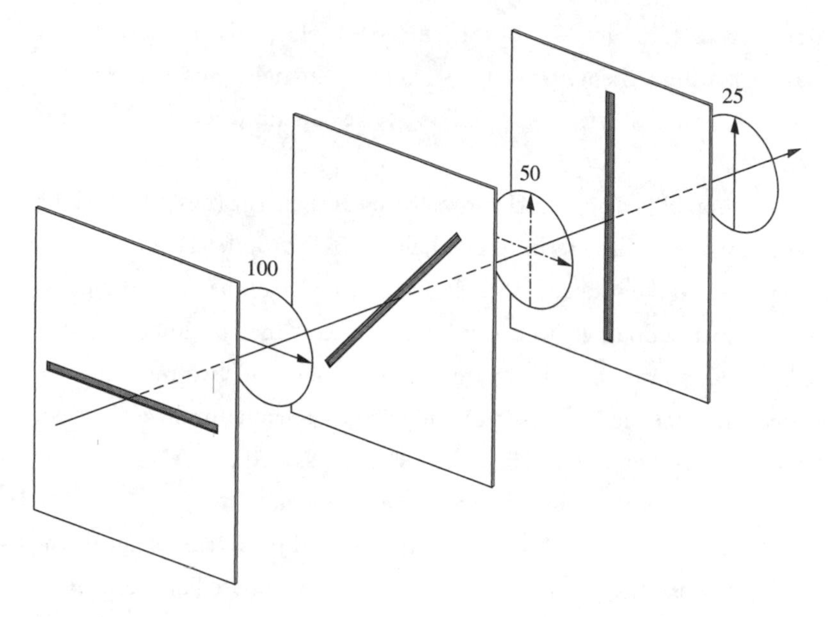

Figure 3.3. Three-way polarizers.

bility of being either horizontal or vertical when measured, so some can still get through the final, vertical polarizer. This bizarre result emphasizes that in quantum physics probabilities are *real.* A photon that has passed through a diagonal polarizer is both horizontally and vertically polarized at the same time with a 50 percent chance of each being the result of measurement.

Once Shimony and Clauser's team had decided to go with light rather than electrons, they still had to produce a steady, controlled supply of photons in the special entangled state. Their experiment used calcium, the element that provides the hardness in bones, limestone, and chalk, as a source of light. Calcium was heated in an oven to produce a stream of energetic calcium gas. As this flowed out of the oven it was hit by a powerful ultraviolet light beam. Some of the energy from the light was absorbed by the electrons in the calcium atoms, pushing them up in energy (this tiny change is another example of that much misused expression we've already met, a "quantum leap").

A short time later, the electrons with extra energy (described by the emotional-sounding term "excited electrons") drop back to their normal state. In the process, most electrons will give off a single photon of light, corresponding in energy to the photon that originally excited them, but a few—around one in a million—give off two photons, one green and one violet. These photon pairs are born naturally entangled, with polarization that will be linked when measured (that's "correlated" in physicist-speak)—linked by the entanglement.

Over the next couple of years, there would be several experiments undertaken using photons, both by Clauser, now based at the University of California–Berkeley, and by Richard Holt, who had stayed up at Harvard. John Bell seemed pleased that his theoretical test was to be undertaken. He wrote to Clauser:

> *In view of the general success of quantum mechanics, it is very hard for me to doubt the outcome of such experiments. However, I would prefer these experiments, in which the crucial concepts are very directly tested, to have been done and on record. Moreover, there is always the slim chance of an unexpected result, which would shake the world!*

If there seems to be a slightly wistful tone to Bell's clear hope of something unexpected turning up (however unlikely that might be) in the letter, it's not surprising. Despite any evidence the contrary, he would have liked it if Einstein had been proved right in his assertion about the incompleteness of quantum theory. Bell made this very clear in an interview with Jeremy Bernstein:

> *For me, it is so reasonable to assume that the photons in those experiments carry with them programs, which have been correlated in advance, telling them how to behave. This is so rational that I think than when Einstein saw that, and the others refused to see it,* he *was the rational man. The other people, although history has justified them, were burying their heads in the sand . . . it is a pity that Einstein's idea doesn't work. The reasonable thing just doesn't work.*

Unfortunately, the experiments weren't as conclusive in their results as he suggests, and it would be a decade later and on a different continent that the matter would finally be established beyond reasonable doubt.

This is a point worth stressing because, unlike in the movies, it's the way real science usually happens. The simplified, Hollywood picture of scientific endeavor wants decisive breakthroughs that prove the point conclusively (after struggle, failure, and hopefully some steamy love interest). But often, in reality, there's a painstaking process of accumulating results that may well include contradictory findings along the way. At least one of the early attempts in different laboratories to carry out experiments based on Bell's paper failed to support quantum theory. This doesn't mean that the theory was wrong but simply that the early experiments were pushing the technology available to its limits. The potential errors in the experiments were too large to be absolutely certain of the results.

How such errors can occur with limited results becomes more obvious if we revisit the tossed coin. Normally, you would expect the result to be heads 50 percent of the time. If you tossed a coin twice and got a head both times, the result don't support the 50:50 theory—maybe with this coin you get a head every time—but the amount of potential error in two tosses is very high. Two tosses don't prove anything. If the coin had been tossed two hundred times and came up a head every time, it would be a very different situation. By then, the 50:50 hypothesis would be highly unlikely. Chances are, we would be dealing with a double-headed coin.

In the same way, in the entanglement experiment, a single conflicting result was too small a sample to disprove the theory, especially bearing in mind how close the experimenters were to the measurement limits of their equipment. What scientists should do (and did do) when working near the limits of accuracy is not to ignore the unexpected result, but to make sure that the experiment is repeated enough that oddities are highlighted as errors.

As John Waller (lecturer in the history of medicine and biology at the Centre for the Study of Health and Society and the Department of History and Philosophy of Science, in the University of Melbourne, a contender for the longest job title in science) points out in his book, *Leaps in the Dark,* when you are working at the leading edge of science it is all too easy for experiments to be misleading. Waller remarks, "The problem here is that novel experimental protocols involve high degrees of error and throw out lots of rogue results that have nothing to do with nature and everything to do with faults in the experimental setup."

Waller goes on to describe how Isaac Newton's famous experiment that showed white light was made of a mix of the colors of the rainbow, the experiment that Newton called his *experimentum crucis* (crucial experiment), actually failed to support Newton's premise. It did produce the desired results *sometimes,* but many attempts to reproduce the experiment didn't come up with the same effect. The reason was that the experiment depended on using high-quality glass prisms to split and recombine the light—a technology that was not readily available in Newton's day. He admits himself that he bought his first prism at the Stourbridge Fair, an annual event on the banks of the river Cam between the villages of Chesterton and Fen Ditton. At the time, prisms were toys and baubles, not precision scientific instruments, and the experimental results produced with them were unpredictably variable.

Exactly the same thing happened in the first attempts to test out the result of Bell's theorem. The early experimenters were working so close to the limits of their equipment that the results were shaky. But typically this uncertainty decreases with time. Within fifty years of Newton's first manipulations of a beam of sunlight passing through a hole in his window blind, much better optical prisms were available that demonstrated the results—Newton's predicted results—with absolute clarity. In the case of entanglement it would take only a few years, and a young French scientist working in a basement, to push experimental accuracy beyond the bounds of reasonable doubt.

When Alain Aspect returned home from Africa, he set up shop in an underground laboratory in the Center for Optical Research, at the University of Paris. Aware of the inconclusive nature of previous experiments, he was determined to reduce the errors involved to a minimum and to avoid any potential for misinterpretation. One essential for getting it right was to really know the equipment—to be sure of this, Aspect constructed his equipment himself, instead of relying on the technicians who might normally be expected to fabricate the basic components.

Like Clauser, Shimony, and their colleagues, Aspect needed entangled photons. His were also produced from calcium atoms but, in this case, atoms that were hit by twin laser beams. It was the use of modern laser technology, still something of a novelty for Aspect, that made it possible to produce a much more definite result than the earlier experiments. Even so, the rig wasn't ideal. As Alain Aspect was later to comment:

> *The ideal source would be one atom of calcium: we would excite this atom of calcium in a particular way, and then observe the light—a pair of photons—emitted by the atom as it gives up its energy and drops back into its normal unexcited state. In fact it's not as simple as that because we cannot trap and control a single atom of calcium so precisely. Instead, we have an atomic beam—a collection of atoms traveling in a vacuum chamber . . .*

Producing many more entangled photons thanks to the controlled power of his lasers, Aspect was able to get clearer results than his predecessors, and, in a variant on the experiment, he also managed, quite literally, to add a further twist. The trouble with trying to pin down the elusive action of entanglement is that there is always the suspicion that the results could be misleading. Perhaps, for instance, there was some way that the detectors measuring the photons' polarization were conspiring to fix the results. There was a small but real possibility that the detectors, an essential part of the experimental setup, were sharing information through some unknown mechanism using (say) the wiring of the experiment rather than the spooky linkage of entanglement.

With a spark of genius, Aspect devised equipment based on a suggestion made many years before by David Bohm. The idea was to make it impossible for the detectors to conspire by changing their orientation while the photons were still in flight. Bear in mind that the important aspect of the experiment was comparing the results for detectors set at different angles to each other. If each detector knew the angle the other detector was set at it could, by some unknown means, produce an appropriately correlated result. (Of course, it couldn't literally know about the setting, but one detector could in some way be influenced by the orientation of the other.)

Aspect managed to get his detectors to flip direction millions of times each second. Even in the small time the photons took to travel from the calcium source to the two detectors, the orientation of the detectors would have changed, so there was no opportunity for collusion. This was no trivial task. With the technology he had available it was physically impossible to move the detectors mechanically at this speed, but Aspect knew something surprising about water that he could use to his benefit. Water's refractive index changes when you squeeze it.

The refractive index of transparent substances is one of those aspects of high school science you're probably glad you don't have to bother with anymore—but it provides the experimental physicist with a valuable resource. The refractive index describes the way a material bends a beam of light when the light passes into the substance or out of it. Refraction is the effect that produces the spectrum of colors from a prism (because different colors bend by different amounts), and makes a coin hidden at the bottom of a cup move into sight as the light from the coin bends when water is poured in; the refractive index measures how strong that bending effect is.

For any particular refractive index, there is one angle (the critical angle) at which total internal reflection occurs. If a beam of light hits the edge of the substance at this angle to the perpendicular, or at a

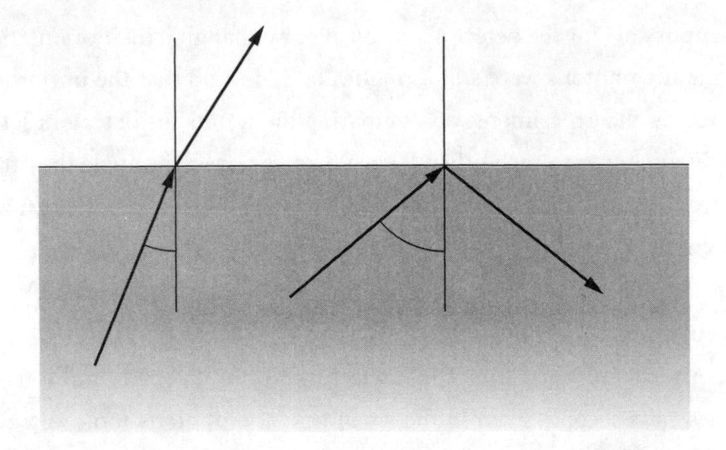

Figure 3.4. Refraction and total internal reflection when light passes from water to air.

greater angle, the light fails to pass through, reflecting back into the substance.

Take light that is passing through water and out into the air. If the light hits the water surface at an angle close to the perpendicular, it will continue on, bending away from the vertical. But if that light hits the surface of the water at greater than the critical angle from the perpendicular—in the case of water to air, 48.6 degrees—it will bounce back.

What Aspect knew, making refraction so valuable, is that the refractive index of water increases slightly when the water is compressed. If photons are sent through the water at close to the critical angle and a pressure wave is set up by squeezing and relaxing the container very quickly, the same way that a loudspeaker cone sets up sound waves by rapidly moving back and forth, the critical angle will change just enough during the "squeeze" part of the cycle to reflect the photons instead of letting them pass through. In Aspect's experiment, this happened 25 million times a second. A pair of devices called *transducers* that rapidly change shape when an electrical current is passed through them, repeatedly squashed and released the water container to set up the change.

The water acted like a railroad switch, diverting the light one way or another. If the photon happened to hit the water/air boundary when the water wasn't under pressure, it passed straight through and hit a first polarizer. If the water was under pressure as the photon hit the boundary, the light was reflected back through the water and into a second polarizer, set at a different angle to the first. The direction of polarization was switched 25 million times a second. Technically this isn't a random change as the variation was regular, but there were was no connection between the timing of the random photon discharges and the timing of the polarizer changes. And those changes were being made too fast to enable any "conspiracy" information to flow between the two ends of the apparatus, near 50 feet apart from each other.

What Aspect's experiments showed is that, beyond reasonable doubt, Bell's theorem provided a successful confirmation of quantum theory. The experiment came up with the results predicted by the theory, demonstrating that Einstein's assumption, that local reality always holds, is false. The phenomenon Einstein considered spooky action at a distance was real, not the unacceptable outcome in a thought experiment.

Even with the success of Aspect's experiment, a few diehards argued that entanglement's instant connection hadn't been proved. It could be, some argued, a little like the disruption of simultaneity that occurs with special relativity (see chapter 5). The change, they believed, only *seems* instantaneous when seen from one end of the experiment. Looking from the other end, it was only after a checking piece of information had traveled the length of the experiment at light speed that the Bell's inequality showed up. But most saw this argument as clutching at straws—Aspect's experiment and many more since have every appearance of generating this instantaneous response.

Aspect was asked what he thought Einstein would have made of the results of his experiment, had he been alive. He responded with typical Gallic charm:

> *Oh, of course I cannot answer this question, but what I am sure of is that Einstein would certainly have had something very clever to say about it.*

There were two reasons why it took so long to get from John Bell's theoretical paper to a definitive experimental demonstration of entanglement's power. The first was the lack of enthusiasm for investigating quantum theory in the late sixties and early seventies. The other was those infuriatingly elusive photons.

Although lasers had made the production of entangled photons much more reliable, entangled pairs were still difficult to come by. Even worse, the direction in which the photons flew off from the calcium atoms was not always the same. It was a tricky business, ensuring that the two photons that were compared really were the entangled pair, rather than accidentally matching up an arbitrary, unconnected couple of photons.

With a renewed interest in the whole concept of entanglement, a major effort was put into devising a more reliable source of these strange, interconnected twin particles of light. The answer came from a freak of nature, a phenomenon that revels in the mind-numbing name of Spontaneous Parametric Down-Conversion. It's a process that relies on the behavior of a crystal.

Scientists are as enthusiastic about crystals as New Agers, though for very different reasons. Science doesn't ascribe arcane powers and mystical energies to crystals, but natural crystals have played a very important part in understanding some of the fundamental phenomena of nature. The way X-rays were bent by crystals was essential to the understanding of the composition of materials—and led to discoveries like the structure of DNA—while the very odd behavior of light in a crystal called Iceland Spar was the inspiration behind the discovery of polarization over three hundred years ago.

A typical piece of Iceland Spar, a form of the mineral calcite, looks like an irregular chunk of very poorly made glass but, when light passes through the crystal, it is split into two. Put a block of the spar on a page of this book and you will see two copies of the words below, seeming to float mysteriously in the crystal. When Erasmus Bartholin, an enthusiastic Scandinavian experimenter, first described this effect in 1669 in the snappily titled *Experimentia Crystalli Islandici Disdiaclastici,* he believed he had discovered something new and fundamental about light. There was, thought Bartholin, not one type of light, but two, identical in appearance but differing in behavior.

We now know that he was wrong, although his discovery was still important. The crystal splits light into two polarizations. It is a natural Polaroid material, but instead of excluding some polarizations it shifts the light sideways according to its polarization, producing a second image. Strange though this behavior is, it pales by comparison with the effect that would make production of entangled photons a more practical, everyday laboratory feat.

Shine a laser beam through another type of crystal and the main beam becomes surrounded by a halo of shimmering colors. It's like something out of an Indiana Jones movie. These special crystals (barium borate and lithium iodate are two of the best-known examples) are doing something very strange as the light is transmitted through them.

To understand what is happening, we need to take a detour into the unlikely world of QED, quantum electrodynamics. This is the aspect of science most frequently associated with the unrivaled U.S. physicist Richard Feynman—it describes how light and matter interact, and as such is one of the most important processes in all of science. QED tells us that the mental picture most of us have of light's interaction with atoms of matter is hopelessly oversimplified.

Look in a mirror. When you see your reflection, light has already traveled from some source (the sun, a lightbulb, whatever), hit your face, and then headed off toward the mirror to collide with the shiny mirrored surface. According to science as we were taught it at school, the light bounces off the mirror and comes back toward your face, ending up being focused on the retina by the eye, where it produces an image that can be interpreted by your brain.

Take another look at what is happening, but think in terms of photons, imagining light as a stream of tiny pellets of energy, hurtling toward the mirror. Our high school science tells us that the minuscule bullets of light bounce off the mirror like pool balls bouncing off a cushion. Only that's not really what happens, something that becomes obvious if you think yourself down to the scale of the tiny particles we are dealing with.

Along comes a photon of light. It's incredibly small and insubstantial. You can't feel it or weigh it or touch it. It gets closer and closer to the mirror. (For simplicity, let's imagine this is a mirror without glass, just a polished metal surface.) The mirror itself is, of course, made up of atoms. Each one of these atoms consists of a vast open space with a few widely distributed components. Somewhere near the middle of the atom is the nucleus, the heavy part. This is so small compared to the space the atom takes up that in the early days of atomic physics, the nucleus was often compared to a fly lurking somewhere in the heart of a cathedral. (Ernest Rutherford, the man who discovered the basic atomic structure, preferred to describe it as a gnat in London's Albert Hall.) That's how much empty space there is in the atom.

Somewhere around the outside of that space are the electrons, much, much smaller than the nucleus. We know they're moving, but not exactly how or where. It's best to think of them as a spread-out fog of motion, with all the fuzziness of the Roadrunner at speed in the animated cartoons, rather than the old idea of planets stolidly circling the sun.

So what does our photon do? What exactly is it going to bounce off, in good pool-ball fashion, when it hits the mirror? The nucleus? If that was the case, the vast majority of photons would pass straight through. Every material would be transparent with just a tiny fraction of the light bouncing back. So if it's not the nucleus, could photons bounce off that vague cloud of electrons? Well, possibly, but why should something as insubstantial as a photon behave like a pool ball?

In fact, light doesn't bounce off the mirror at all. What happens is much more interesting. Light is electromagnetic energy. As its electric field gets close to an electron, they begin to interact. The electron absorbs the photon of light, eats it up, becoming more energetic in the process. It takes a quantum leap. But electrons in the higher energy states are often unstable. Soon it pulses out a new photon of light. The photon shoots back from the mirror surface toward your eye. The "reflected" photon is actually a totally different photon from the incoming one.

Similarly, when light passes through a transparent substance, all the photons don't shoot straight through it: instead they get absorbed and reemitted as they go. (If all the photons did pass straight through without interacting, there would be no difference between light's behavior in a transparent material and a vacuum.) This happens with all transparent substances, but in special crystals such as barium borate, an extra phenomenon occurs.

When laser light with enough intensity hits such a crystal, some of the photons that are absorbed result in not one, but two photons being reemitted. The amount of energy in a light photon depends on its frequency—light's equivalent of the pitch of a musical note. The higher the frequency, the higher the energy, from low-frequency photons like radio, through microwaves, infrared, visible light from red to violet and on up through ultraviolet, X-rays, and the rest. When two photons are produced, the two new photons have a total frequency adding up to the frequency of the original light, so no energy is lost. The "down" in this "down-conversion" process refers to the fact that for any given photon the frequency goes down compared with the original. Just like the pair of photons emitted by the laser-stimulated calcium, the photons here are entangled, but there are three big advantages to using this technique.

First, although the photon pairs are still in the distinct minority, they are much more abundant from this process than the entangled photons produced by calcium. Second, it's a less volatile physical environment. Instead of dealing with a tricky stream of hot calcium atoms, the researcher has a nice, solid, well-behaved crystal to deal with. Finally, and most important, there's an extra benefit. Not only are the two photons entangled, but the directions in which they emerge from the crystal are more predictable. If one entangled photon goes off in direction A, the other will always go off in direction B—making it much easier to be sure just which ones *are* the entangled photons, always a problem with the calcium method.

With more secure results from down conversion proving even more conclusively that Aspect's experiments had delivered the right result, it might seem there was little need to take Bell's theorem any further. Yet there was always a slight discomfort from the way it was proved using inequalities. All the experiments so far had a measured a set of values, and if they fell statistically outside a range determined by John Bell's calculations, then Bell's theorem and quantum theory were assumed to be true and local reality was dismissed.

The outcome would feel more conclusive if it were possible to prove the result in a more positive way, rather than showing that, averaged across many readings, something *didn't* happen—and this was to be the role of the second three-letter acronym that is familiar to those who work in the world of entanglement. What was probably the most definitive answer to EPR came from GHZ.

The GHZ experiment (we'll come back to why it's GHZ in a moment) entangled not two, but three photons. At first sight this seems like an embarrassing case of "anything you can do, I can do better." And why stop at three? There's surely a dangerous encouragement to escalation, entangling more and more particles, without advancing knowledge in any way. Yet it happens that the three-photon experiment (which was developed theoretically in the mid 1980s and was finally undertaken in 1999) had one huge advantage over its predecessor.

Where Aspect's experiments, and all the others of that period, worked by counting up the statistical results of the *partial* correlation from many, many photon pairs—and proved Bell's theorem when on average these strayed beyond a certain limit—the new, three-photon, experiment had an absolute result. Each measurement either supported or countered Bell's results. There was no uncomfortable statistical feel to the outcome of the experiment.

This three-particle approach was the work of Mike Horne, the experimenter who helped Shimony, Holt, and Clauser, along with American physicist Danny Greenberger and Austrian quantum expert Anton Zeilinger. (When the paper was written up, Abner Shimony also contributed as one of the leading EPR gurus of the time, but the initials GHZ still stuck.) The experiment produces two pairs of entangled photons, which then become linked together as a triplet when one of the four is captured. The math to produce the result is messy and really isn't necessary to understand what is happening. The important outcome is that Bell's theorem can be proved with a positive, actual measurement, rather than an indirect implication. And when the experiment was finally carried out in 1999 by Anton Zeilinger and his team, the result was positive. Quantum theory was vindicated and the local reality that Einstein and colleagues had struggled to protect in their EPR paper was conclusively shown to be impossible.

Up to this point it might have been possible to put off facing the true existence of entanglement as anything more than a paradoxical theory, but the experimental evidence by now was overwhelming. This, then, is a good place to explore a concern that many feel when they hear about entanglement. It was recently voiced in a letter to the magazine *New Scientist* in which the author wondered what all the fuss was about. When particles are entangled, he assumed, their state (spin, polarization, or whatever) is known, so there is no problem with them having linked values, because they had these values ever since entanglement. John Bell himself illustrated this viewpoint delightful with the strange case of Doctor Bertlmann's socks.

One of Bell's lesser claims to fame was publishing a paper called "Bertlmann's Socks and the Nature of Reality," surely one of the strangest titles in all of science, in the *Journal de Physique.* (Reinhold Bertlmann was another scientist who worked closely with Bell.) The Doctor Bertlmann in question has unusual taste in clothing. He chooses the color of his socks randomly, but always makes sure that he puts on socks of different color. He never wears a matching pair. Because of this, when we see Dr. Bertlmann appearing around the corner of a building and catch a glimpse of his left foot, noticing a pink sock (as Bell wryly comments, there is no accounting for taste), we can be sure that his other sock is not pink, even though we have never seen it.

Unlikely though it may seem, this wasn't a matter of John Bell coming up with an imaginary example to entertain as it explained. (Bell rightly seemed to think that most scientific papers were unnecessarily dull, and his were significantly more readable than most.) In fact, Reinhold Bertlmann really did always wear odd socks. This was something he began doing in his late teens—according to Anton Zeilinger, who was later a colleague, "as a protest against the establishment"—and he had continued to do so ever since.

To stretch the analogy and make it even closer to the claims for entanglement in a way that Bell was too tasteful to do, had Bertlmann suffered a terrible accident shortly after getting dressed in which his right foot was severed, and had he then hopped into a spaceship and crossed to the opposite side of the universe from the Earth, as soon as we saw the unipodal Bertlmann with a pink sock, we would instantly know that the other sock wasn't pink, despite the huge distance separating us from it. Even though the information that told us what color the other sock was would take millions of years to reach us. Spooky. Or rather, not spooky—so what is all the fuss about?

In fact there is a fundamental difference between Bertlmann's socks (or the world as seen by the writer of the letter to *New Scientist*) and what actually happens in quantum entanglement. With the socks, the color is already fixed when Bertlmann puts them on. The decision is made. The information (second sock set to different color—green, say) is inserted into the system at that point. If we were dealing with quantum socks, however, the state being observed (spin, polarization, or whatever) is not fixed at the point the entanglement takes place.

Remember the experiment with the three polarizing filters. It wouldn't work if the photons passing through the diagonal filter had just maintained their fixed horizontal polarization. If Bertlmann's socks were quantum particles, both socks would have all possible colors until we saw one of them. At the moment that sock, inspired by nothing more than random chance, settles on pink as the color, and only then, the other sock fixes its color to something else.

The experiments derived from Bell's theorem proved that there are no local "hidden variables." There was no hidden value of spin or polarization set at the point at which entanglement was established, ready to be observed after the particles were separated. If it seems I'm overemphasizing the point, it's because so many people still fall back on the Bertlmann's socks position—but things really aren't like that. The Bell's theorem experiments distinguish between a preset value and one that was randomly attributed when the observation was made—and it's the preset value that had to be thrown out the window, much to the reluctance of many involved.

The spooky action of entanglement is fascinating in its own right, but, as will become obvious in the next chapter, it has the potential to deliver otherwise impossible feats of communication, computing, and matter transmission. If entanglement is to be of any use in the real world, though, it's not enough to be able to make entangled particles in the lab. It has to be possible to send these particles across a significant distance, or to store them for a reasonable amount of time, while keeping them entangled. It's all very well to say that you could separate entangled photons to opposite sides of the universe and a measurement of the polarization of one will have instant implications for the other—but that's not a lot of use if you can't send a photon more than a few yards without the entanglement collapsing.

When Schrödinger first conceived of entanglement, he assumed that it would only be possible across distances where the travel time of light was smaller than any significant processes in the system (so, he believed, entanglement could never be put to practical use). Initially, even among active experimenters there were concerns that the scale of the laboratory was the limit of keeping particles entangled. But since the turn of twenty-first century, there has been a huge increase in experimenters' ability to maintain entanglement under more and more robust conditions.

The earliest successes came in sending entangled light down optical fibers, the sort of communication network that provides the backbone of the worldwide phone system and the Internet. This long-distance entanglement through hair-fine glass threads is effective, but limited. All signals in fiber-optic cables lose strength as the photons interact with the walls of the fiber. A typical commercial cable will not send a signal further than 40 miles without regenerating it, using a special amplifier called a repeater. So far, the farthest entangled photons have been sent is around 100 kilometers (62 miles), but the losses at this distance are huge, and a more practical distance is 20 kilometers (12½ miles).

By the standards of worldwide communication this seems an impractically short range—but the early developers of the telegraph hit similar problems. Just as they used different ways to repeat the message, sending it on down the line every few miles, and the optical fibers that span the globe use amplifying repeaters, so Peter Zoller and his colleagues at the University of Innsbruck, in Austria, have come up with a technique to establish a repeater technology for entanglement.

Using Zoller's idea, Alex Kuzmich and Dmitry Matsukevich, of the (Atlanta) Georgia Institute of Technology, managed to entangle two clouds of rubidium atoms, cooled to near absolute zero. In this state, the cloud of atoms acts as a single entity. When it emits a light photon, this photon is entangled with the entire cloud. A photon from each cloud was sent through a device called a beam splitter. This sounds like some high-tech piece of science fiction equipment, but a good example of a beam splitter is the half-silvered (or one-way) mirror, popular in TV cop shows for secretly watching interrogations. It's a device that lets part of a beam of light through and deflects another part in a different direction.

Don't be fooled by the apparent simplicity, though: at the quantum level, beam splitting is a weird business, hence its ability to entangle photons. To better understand what's going on, let's think a little about a very simple kind of beam splitter that we've all got in the home—a regular glass window. Stand in front of a window in a well-lit room at night and look at the glass. What do you see? Yourself. The window has turned into a mirror. If it were an ordinary mirror, nothing would come out the other side. But if you were to go outside your house and take a look at that same, reflecting window, you would clearly see the well-lit room. So a fair amount of light—in fact most of it—is passing through. The window of your home has turned into a beam splitter. (In truth, it always was—some light is always reflected back—it's just less visible in the daylight.)

We accept this partial reflection as common sense and natural, but once you start to think about what's happening, it's anything but an obvious occurrence. Let's imagine the light from the room as a stream of photons, hitting the surface of the glass. Some of those photons will be reflected back. Most won't. So how does a particular photon know what to do? It's the usual quantum challenge—we know the probability that something will happen, and on average the right quantity of photons will be reflected, but what makes one photon travel through and another bounce back is a mystery.

This was a problem that Isaac Newton faced as well. He was convinced that light was made up of corpuscles, tiny particles, and couldn't understand how some light hitting a window bounced back and some didn't. The most obvious suggestion was that there were irregularities in the surface of the glass. If this were the cause, it would be as if tiny bits of the surface were mirrors, while other, larger areas were clear. So corpuscles hitting the mirrored segments would bounce back, while the rest traveled through. It makes a lot of sense. But, as Newton realized, it's wrong.

Newton had done a fair amount of lens making for his optical experiments, and he knew that as you polish the glass of a lens with finer and finer material, resulting in smaller and smaller scratches, it becomes transparent. Very fine scratches don't seem to affect transparency—yet if the cause of partial reflection were irregularities in the surface of the glass, these bumps and cavities would have to be *very* small, so small that they could not be seen, and so the glass shouldn't be capable of producing those ghostly images.

With a modern quantum viewpoint, there is no obvious cause for the partial reflection. As with so many other aspects of quantum theory, we simply have to accept the probabilistic nature of the process (even though, like Einstein's, our minds rebel against it). But this is only the start of the strangeness of the beam splitting window—because light passing through a piece of glass hits not one, but two interfaces. First, it passes from air to glass as it leaves the room, then it moves from glass to air as it escapes outside.

It's not surprising that the second transition can also result in some reflection—but things aren't that simple. The amount of partial reflection varies depending on how thick the glass is. "So what?" you are probably thinking. It should hardly be surprising that the amount of reflection from the outside of your window depends on how thick the glass is. But in practice the amount of reflection from *both* surfaces of the glass depends on the thickness. With certain thicknesses you can reduce the reflection from the *inside* of the window to practically nothing.

Think about that for a moment. You shine a beam of light on the inside of your window. Normally, some photons will reflect back. But if the glass is the right thickness, they will somehow know how thick the glass is at the point they hit the *inside* surface of the window, and instead of reflecting back, they will carry on through. Strange indeed.

Once you realize just how bizarre the process of partial reflection is, it's somehow not entirely surprising that a beam splitter can produce entanglement. The method used is a little complex, but not too painful, taken step by step. The process starts by firing a photon at a beam splitter from which it could emerge, say, to the left or the right. We make no measurements, so this being a quantum process, the photon doesn't magically choose one path or the other—it is in a superposition of two states, one where it flew off to the left, one to the right.

Waiting in the path of each of these potential photons is one of the two cold clouds of rubidium atoms. If a photon hits the cloud it might be absorbed, then a new photon is emitted, leaving the cloud in a slightly different, more energetic state. The reemitted photon (whichever cloud it hit) goes into another, special beam splitter called a polarizing beam splitter. This reflects or transmits depending on the polarization of the photon. At this point we measure the state of the photon, and the outcome of the act of measurement is to force the two clouds into entanglement (it doesn't matter which of the possible output states the photon is in: either implies that the gas clouds will be entangled). The outputs from the second beam splitter are associated with a superposition of the possible input states. That's a superposition of an interaction with each rubidium cloud—and that means as we make the measurement it pulls together a mix of the states of the clouds, dragging them into entanglement.

That is as far as the experiment has gone to date, ending up with two clouds that, in principle, could be up to 40 kilometers apart (25 miles), each 20 kilometers down a fiber-optic cable from the point where the original photon passes through the beam splitter. But the idea is that this entangled link could ripple on outwards. Each cloud could then be entangled with another cloud a further 40 kilometers away, ending up with entangled clouds (which could then act as sources of entangled photons) as far apart from each other as is desired. "One goal," Kuzmich has said, "is to build a long-distance quantum connection between, say, Washington, D.C., and New York City."

In 2003, entanglement also burst out of fiber optics, an essential if it is ever to be used in the other mainstay of long-distance communication, satellite technology. Researchers at the University of Vienna managed to send entangled photons from one side of the river Danube to the other. This was no cushy lab job. The initial experiments had to be undertaken at nighttime, because the beam of entangled photons would be lost among the swarm of natural photons from the sun. In freezing cold weather and howling winds, the team set up miniature entangled photon generators by the river, housed in a rusty old shipping container. These pumped their light through a focusing telescope, to be picked up by another telescope across the water. The telescopes were hand built for the experiment. (There was nothing significant about crossing the Danube, it just made the result sound more newsworthy.)

The entangled photons crossed the 600-meter (2,000-foot) gap and were still safely entangled when received. Next step was to cover an even greater distance. Rival projects in China and Austria have been attempting this. First to publish was Jian-Wei Pan's team from the University of Science and Technology of China, in Heifei. In 2004, they managed to send a beam across two links, one 7.7 kilometers (4¾ miles), the other 5.3 (3¼ miles). Their next experiment is expected to rival Anton Zeilinger for showmanship. The Chinese team plans to link two stations 20 kilometers (12½ miles) apart on China's most famous monument, the Great Wall. "We used to light fires on the Great Wall to signal invasions," commented Pan. "Now we are going to signal the future."

Meanwhile, the latest outdoor experiments from the Austrian team covers two legs of 7.2 and 8 kilometers (each nearly 5 miles) in length. The transmitting station is on top of a small astronomical observatory that nestles in the hills above Vienna, with receivers in two of the city's skyscrapers to enable a clear line of sight. After the experience of trying to set up delicate equipment in the freezing cold Vienna night, the decision was made to put at least some of this experiment inside the buildings, though this attempt to work in comfort nearly wrecked the whole exercise.

Like many modern glazed structures, the VTT, one of the two office buildings in the experiment, had special window panels that reduce the transmission of infrared. This has the double bonus of cutting down on heat losses from the building, and stopping the office from turning into a greenhouse on sunny days. Unfortunately that same coating totally blocked the entangled photons. Nothing was getting through. It was lucky for the team that the enlightened owners of the building were prepared to replace the window in the office housing the experiment with conventional glass to enable the transmission to go ahead without further risk of frostbite. In the other building, the Millennium Tower, the researchers were given half of the whole top floor of the building to construct their experiment—an unusual luxury for a scientific team.

The distances covered in this experiment and its Chinese counterpart were not arbitrary, but chosen for a very significant reason. It turns out that sending a beam of photons through around 6 kilometers (3¾ miles) of air at ground level provides the same resistance to progress as transmitting a signal all the way from the ground to a communication satellite thousands of miles above, and satellites are essential for long-range communication, particularly if entangled photons are to be sent a good distance without entanglement repeaters.

The reason there is such a disparity between the distance at ground level and the range heading into space is due to the rapid thinning of the atmosphere as you head away from the Earth. Even at a mere 8,000 feet, around the altitude of Mexico City, the height equivalent of the minimum pressure allowed in the cabin of a passenger airliner, there are 20 percent less gas molecules around to scatter the photons and upset transmission than at sea level. By the time you reach 68,000 feet (around 13 miles), there is hardly any air.

Even though a satellite might operate 22,000 miles up in space, the entangled beam will only have to fight against the equivalent amount of air as it would traveling around 3¾ miles at ground level. The successes of the long-range experiments in Heifei and Vienna show that it is possible to use a satellite to provide entangled photons, a highly intriguing prospect, as we shall see in the next chapter.

These achievements were made in 2004, though the realities of working outside the lab meant that it proved harder than expected. Air pollution and fluctuations in the atmosphere from the heat generated by buildings, cars, and people made it difficult to keep the beams in communication. At any one time, the Austrian team only managed to get one of the links going. To activate both legs, they have had to introduce adaptive optics.

Imagine looking down a long, straight road on a hot day. The surface seems to shimmer, turning to liquid—the air above the road dances and ripples. Adaptive optics are designed to undo such warping, stabilizing the image and removing the complex distortions that make it so difficult to keep a light beam static over a city.

Before powerful computers were available this simply was not possible, but adaptive optics uses a range of computer controlled technology to clarify the image. In some equipment, the light is first bounced off a conventional mirror that can be very quickly steered to change the angle of tilt. This irons out vibrations. The key part of the system, though, is when the light hits a second, flexible mirror that can be distorted in shape. A sensor samples part of the light before it hits the mirror, and monitors the position of a set of easily identified points—as these move, so the mirror is distorted to undo the movement. Typically the mirror might be reshaped several hundred times a second. "This was great," commented Anton Zeilinger with typical boyish enthusiasm on the opportunity to use the adaptive optics technology, "we always enjoy it when we have to use new toys."

Adaptive optics will also be necessary for the third phase of experiments that the Vienna team has already planned. Where Pan is moving to the Great Wall of China, Zeilinger intends to head for space. The next stage will use a specialist telescope on Tenerife, one of the Spanish-owned Canary Islands located off the coast of North Africa. The telescope can act as beaming device to provide an optical link to a satellite high above.

The European North Observatory, on Tenerife, is a startling sight. After crossing mile after mile of barren, volcanic landscape, a primeval island terrain littered with ominous, cinder black, jagged rocks, the domes of the observatory suddenly appear, perched 2,500 meters above sea level. The white, pure shapes of the buildings seem like a futuristic outpost established on another planet. And it's from here that the entangled photons will be beamed up to a satellite and back to test the concept of distributing an entangled communication channel through space. The experiment can't be undertaken until there is a suitable receiver in orbit—Zeilinger hopes that it will be launched by 2010.

This all sounds like a perfect, step-by-step development from the tightly controlled lab environment to the real world, gradually achieving more and more. But as researchers stretch the practical application of entanglement over greater and greater distances, they will have to face another problem, over and above the difficulty of getting the photons securely to the right place. They will have to deal with one particular and deceptively trivial-seeming challenge—making sure that they know which way is up.

This apparently innocent problem is one that has been highlighted by Australian physicist Howard Wiseman, of Griffith University, in Queensland. Wiseman, acting as a sober pessimist in contrast to the enthusiasm of researchers like Zeilinger, has pointed out that there could be a disconcerting distance between theory and reality, so much so that he uses the insulting term "fluffy bunnies" (okay, mildly insulting) to describe entanglements that are great in theory but can't be used in practice.

What's the problem? A lot of the measurements used in entanglement involve a direction. If dealing with spin, for instance, we need to know the direction in which spin is being measured. Similarly, for polarization of photons, the two ends of the communication channel need to swap details of the polarization direction they're measuring for the photons. That's fine in the lab, where up and down have a simple, consistent meaning, but as a transmitter and receiver get farther and farther apart, the definition of up or down (or whatever direction you like) becomes fuzzy.

At the extreme, imagine someone at the North Pole setting up an entanglement with someone at the South Pole. "What do you mean by 'up'?" we ask the North Pole scientist. "North, of course," she replies.

"But that's down," says her Antarctic equivalent. Of course, in this simple case, it's easy to correct for exactly what is meant by up and down, but in most cases the position is more complex.

The more uncertainty there is between the two locations in the idea of what is up and down, the more errors there will be in the communication. Things get even more complicated if one location is moving with respect to the other—say, if the receiver is in a plane. Luckily, this isn't going to prove too much of a problem for those looking to make use of entanglement in the real world—they just have to be ready for the problem. In effect, it's a problem of relativity, and the answer is to make the measurements relativistic too.

Instead of looking to measure in any particular direction, relative measurements can be made between two or more particles, throwing away the need to know absolute orientation. In March 2004, researchers at Oxford and Warsaw used twin sets of photon pairs for measurement and managed to get significantly better results than with experiments relying on differentiating up and down. Another team at the Max Planck Institute in Garching, Germany, went even further, using four sets of photons to make relative measurements on. They found that even after passing the photons through a randomizer it was possible to make use of them, because the relative directions of the measurements were still meaningful even though any idea of specific direction was now totally lost.

What has been found means that it may take more quantum particles than had been hoped to transfer the effect of the spooky connection of entanglement from one place to another, but it is still practically possible. And entertainingly, entanglement itself has managed to provide a solution to another potential problem for users of entangled photons—making sure that clocks are synchronized at different locations.

In 2004, Alejandra Valencia and her team at the University of

Maryland used entangled photons to bring clocks into near-perfect synchronization. Entangled photon pairs were sent off from a central source to receivers 1.5 kilometers (9⁄10 mile) away in each direction (the two receivers were 3 kilometers [1 4⁄5 mile] apart). Timings were compared for photon pairs that show up as entangled and the two clocks were gradually brought into synchronization so that the timings coincided. Here, the entanglement was being used as a marker to determine which photons left the central source at exactly the same time. In the experiment, the clocks were brought into line with picosecond accuracy—that's to the nearest trillionth of a second.

With robust transmission of entangled photons possible, the applications of entanglement are no longer restricted to the laboratory. Out in the big, wide world, they are ready to make a real impact on everyday life. And the first commercial use of entanglement is likely to be all about keeping secrets.

CHAPTER FOUR

A TANGLE OF SECRETS

For secrets are edged tools,
And must be kept from children and fools.

—JOHN DRYDEN, *"Sir Martin Mar-All"*

Entanglement provides a secret link, an unfathomable bond between two particles. It perhaps isn't surprising that once practical entanglement became a reality it came to the attention of those whose business is keeping secrets from prying eyes. The initial interest was inspired by the spooky connection between entangled particles itself. If entanglement could be used to carry a message, it would be totally impenetrable to any interception. This was obviously the possibility in the mind of the executive director of a California think tank who wrote to the undersecretary of Defense for Research and Engineering:

> *If in fact we can control the faster-than-light nonlocal effect [entanglement], it would be possible . . . to make an untappable and unjammable command-control-communication system at very high bit rates for use in the submarine fleet. The important point is that since there is no ordinary electromagnetic signal linking the encoder with the decoder in such a hypothetical system, there is nothing for the enemy to tap or jam.*

The think tank director didn't know the half of it—if information could be sent down an entangled link, it would be possible to pass a message back through time. The downfall of that incredible plan will have to wait until the next chapter, but entanglement was to have a more realistic impact on the cryptography trade, one that by 2005 had made its first small inroads into the commercial market.

Communication is the lifeblood of civilization. We are communicating animals—it partly defines the essence of humanity. Yet the urge to communicate carries with it the contradictory need for secrecy. Much communication—broadcast television, say—is intentionally transmitted to the world. We don't care who hears—in fact, the more, the better. Many other conversations are comfortably public, whether it's a shouted discussion across a street or a postcard that's sent openly through the mail. But sometimes we want to get our message to the right person, and only that person.

When the need for security kicks in, we don't want to be overheard. Many conversations require privacy. It's interesting that the telephone as a social vehicle didn't really take off until direct dialing was possible and the operator was removed from the loop. The knowledge that a prying ear could always be eavesdropping made early phone calls seem less personal and reduced their value.

Of course, a normal telephone conversation can still be overheard with the right technology—not too much of a concern for everyday gossiping individuals, but a real threat to those involved in national security or high-stakes business. (Or for that matter, in spying or large-scale crime.) Here, the result of being overheard can mean vast losses, criminal proceedings, or even death or war, rather than a minor social embarrassment. Encryption, keeping messages from eavesdroppers, is essential whether you are a bank wanting to protect electronic fund transfers, a Web site taking online payment, or a military commander ordering an attack.

In the past, spying has been referred to as "the great game," although the term was originally used by Kipling to refer to the tug-of-war between Britain and Russia over control of India. But "the great game" seems even more appropriate for the ongoing battle between cryptography, the science of code making, and cryptanalysis, the equivalent process of decoding secret messages.

The game in cryptography is always on the move. Code makers invent a new method of concealing information. Code breakers find a way to smash through. But by then, the code makers have devised a new strategy—and so it goes on. It really is a matter of mental games. In fact, when Britain was recruiting potential experts for its Bletchley Park codebreaking center during the Second World War, the unit that would crack the immense complexity of the German Enigma encryption machines, one of the criteria used in selection was the ability to solve crossword puzzles.

It's likely that attempts have been made to keep information secret as long as messages have existed. We have to go back a long way in time to find the earliest techniques. The first recorded attempts at secret communication seem to have depended more on hiding the message than making it unreadable. The ancient Greeks were past masters at concealing information. They would scrape the wax off a writing tablet, inscribe the message on the wood below, and replace the wax before sending the apparently blank tablet on its way. A more extreme technique, not exactly practical for an urgent communication, was to shave a messenger's head, write the message on his scalp, and then wait for the hair to regrow and hide the letters.

The Greeks, or more precisely the Spartans, were also responsible for one of the first ciphers, even providing an ingeniously simple device for automatically encoding and decoding the messages. A strip of leather was wound many times around a wooden staff called a *scytale.* The leather was tightly wound, each coil pressed against the next, so that it provided a continuous surface. The message was then written on the leather in lines along the length of the staff (this was easier if the staff had multiple flat sides, rather than being a smooth cylinder). When the strip was unwound, it carried a meaningless string of letters that could only be reinterpreted by fitting the piece of leather around a similar-sized *scytale.*

True ciphers also emerged in the ancient world, most famously in the employ of Julius Caesar, who was an enthusiastic user of basic substitution codes that are still popular in children's magazines. Such codes that consistently replace one letter with another (for example, by shifting a fixed number of letters through the alphabet, or starting at the opposite end so Z represents A, Y represents B, and so on) hide thc message by following very simple rules—ideal if they are to be employed by a soldier in the field.

All of these ancient mechanisms for keeping information secret suffered from one significant drawback. They were easy to break. The hidden messages would only stay hidden while the method was unknown. After all, anyone could wrap a strip of leather around a stick, and a basic substitution cipher like Caesar's hardly challenges the mental capacity of a ten-year-old to crack. Increasingly over the centuries, complexity was added to the cryptographer's armory. It made the codes harder to use but increased the protection against would-be cryptanalysts.

To begin with, substitution ciphers like Caesar's were made more complex by adding extra characters to stand in for special words (common words like "the" and frequently used names). Other extra characters replaced repeated letters (#, for instance, could represent TT), represented spaces, or meant nothing at all, added to the script to cause confusion. Codes were produced where, instead of replacing characters, whole words or phrases were replaced by a single word or symbol. And key-based encryption brought a whole new sophistication to the art of code making.

The idea of a key is very simple, yet very powerful. Key-based encryption systems overcome the biggest snag of any cipher that depends on replacing a letter by another—with a key, a letter in the hidden message doesn't always represent the same letter in the open message. The simplest form of key takes a special word and "adds" the letters of that key word to the letters of the message. For example, if the key word CAT were used to encrypt the message DADDY, I would add three (for the position in the alphabet of the C of cat) to my first letter D to make it G. I would add A (=1) to the A in DADDY to get B. When I come to the second D in DADDY, I add T (=20), turning D into X.

Note that the first D becomes G, but the second D becomes X. Since the key word is shorter than the message to be encoded, we then start over at the beginning of the key word for the fourth letter of the message, encrypting this third D as G again (D+3=G), and the final Y as Z (Y+1=Z). The whole word ends up as GBXGZ.

This ability of a key to confuse by not always using the same character for the same letter in the original message is very powerful, because one of the easiest routes to breaking a cipher is to count the frequency with which each letter turns up in the message. In any language, some letters are used more often than others—in English, for example, E is the most common of all—so provided you have a long enough message, counting how often the different characters occur can give hints as to which character is which. With a good key, those hints lose their value.

A number of factors help make a key even more effective. The longer the key, the better. In my example above, my very short key CAT left two of the Ds represented by the same character, since both the first and third Ds had a C added to them to generate G in the encrypted message. A longer key would overcome that problem. It also doesn't help that my key is, itself, a word that obeys the rules of English—the key could have been improved by selecting random letters, which give nothing away. I could also combine a key with other encryption tricks, like substituting characters for special words and spaces, or having dummy characters that should be ignored.

Provided it is possible to share a key without anyone else knowing what the key is (keeping the key secret can, of course, be as hard as keeping a message from prying eyes), key-based encryption provides very strong security. The famous Enigma machines were based on a sophisticated mechanical key generation mechanism that proved extremely difficult—though not impossible—to crack. This was despite the fact that, as early as 1918, a form of key-based encryption had been developed that was totally impossible to break.

If the key is entirely made up of randomly selected values—so it has no pattern that can be deduced—and if each character in the key is used only once, so that the key is as long as the message to be encoded, the result is totally secure encryption (provided snoopers don't get their hands on the key itself). It's quite surprising that no one came up with this idea sooner, given its relative simplicity, but this is the basis for the "one-time pad."

It was the idea of Gilbert Sanford Vernam, an engineer at AT&T Bell Labs, who suggested that information on a teletype could be safely encrypted by combining it with a previously generated key, kept on paper tape, character by character. (Vernam's original concept was finished off by Captain Joseph Mauborgne of the U.S. Army Signal Corps, who realized that the characters on the key had to be randomly generated to ensure that the message was absolutely secure.) As long as no one else gains access to that key, it is totally impossible to break a one-time pad cipher.

Say I wanted to send a simple message to my broker, "SELL AT 345."

I don't want anyone else to find out and jump into the market before me, so I need to keep the message secret. Just as a computer uses ASCII code, which has a number value for each character, we can apply our own numeric code. Let's keep it nice and simple—the numbers from 1 to 9 represent those digits, 0 is coded as 10, A becomes 11, through to Z becoming 36, and finally a space is 37. We aren't bothering with punctuation, or upper and lower case here, though of course we could use a full-scale ASCII-style code exactly the same way.

This garbled message is sent off to my broker, who also has the private key: 2–31–19–4–16–4–11–27–35–14–9. By reversing the process, the recipient can uncover the message. *Provided that key is never used again*, the technique is unbreakable. The one-time use of the key is essential because otherwise patterns will emerge over a series of messages, giving code breakers a possibility of gradually analyzing the key. But if

Figure 4.1. One-time pad cipher encryption.

<table>
<tr><td>S</td><td>E</td><td>L</td><td>L</td><td></td><td>A</td><td>T</td><td></td><td>3</td><td>4</td><td>5</td></tr>
<tr><td colspan="11">First convert the message into numbers. 1–9 become that number, 0 becomes 10, A is #11 thru to Z becoming 36 and space is 37.</td></tr>
<tr><td>29</td><td>15</td><td>22</td><td>22</td><td>37</td><td>11</td><td>30</td><td>37</td><td>3</td><td>4</td><td>5</td></tr>
<tr><td colspan="11">Next we produce a random series of numbers between 1 and 37 that will be our key.</td></tr>
<tr><td>2</td><td>31</td><td>19</td><td>4</td><td>16</td><td>4</td><td>11</td><td>27</td><td>35</td><td>14</td><td>9</td></tr>
<tr><td colspan="11">Then we add our key numbers to the message</td></tr>
<tr><td>31</td><td>46</td><td>41</td><td>26</td><td>53</td><td>15</td><td>41</td><td>64</td><td>38</td><td>18</td><td>14</td></tr>
<tr><td colspan="11">We can convert those back to characters if we wish by taking 37 off any number bigger than 37.</td></tr>
<tr><td>31</td><td>9</td><td>4</td><td>26</td><td>16</td><td>15</td><td>4</td><td>27</td><td>1</td><td>18</td><td>14</td></tr>
<tr><td>U</td><td>9</td><td>4</td><td>P</td><td>F</td><td>E</td><td>4</td><td>Q</td><td>1</td><td>H</td><td>D</td></tr>
</table>

sender and recipient are the only ones with the key, and only use it once, the one-time pad is totally unbreakable. Without the key, the message is meaningless because the characters have been randomly modified by our random key—there is no way around it, nothing to cling onto when trying to break the code.

Such one-time pads were used in the Second World War, but they could not be deployed in many operational environments because of the difficulty of getting the keys securely to a distant location. If a pad of keys could be intercepted and copied, all the messages encrypted with it would be available to the enemy. It was because of this problem that the Germans introduced Enigma machines. They weren't one-time pad enciphering machines, but produced keys in such a complex way that they were extremely hard to defeat, and the key was produced on the spot, so it couldn't be intercepted.

At the heart of the Enigma device was a series of toothed wheels (initially three, though later machines had more). Each wheel had a series of connectors around the face that touched the adjacent wheel, and each connector was joined within the wheel to another connector on the same wheel's opposite face. When a key was pressed on the Enigma's typewriter keyboard, an electrical current was sent off to the corresponding connector on the first wheel. If that current went to the first connector on the first wheel, it might come out the other side on the fifth connector. This would go into the fifth connector on the second wheel, but come out on the seventeenth connector, say. After passing through the wheels, the current was then "reflected"—sent back through the wheels again, and came back out to light up one of a series of bulbs, one for each connector.

Pressing the A key on the keyboard might result in the S light illuminating. This was the encrypted value. So far this was just a simple substitution, but the Enigma's genius was that when the key was released, the first wheel was rotated one position. Next time, pressing the

A key will resulted in a different encryption. When the first wheel had made a full turn, the second wheel moved on another position—and so on. Provided both ends of a communication channel were using Enigma machines with the same wheels, starting in the same positions (the wheels had letters around the outside so the initial positions could be set), they could exchange messages that used a different cipher connection time after time.

Despite the power of the Enigma machines, and added complexities like extra wheels and a connector board on the front like an old-fashioned telephone switchboard that enabled the link between the keyboard and the first wheel to be changed whenever required, the Enigma messages were eventually cracked by the code breakers at Bletchley Park. Breaking through was a laborious process, depending on a combination of spotting habitual message formats from particular radio operators (one might, for instance, always start with a particular word or a comment on the weather) and brute-force analysis using simple mechanical computers that worked through possible combinations looking for a match. The Enigmas were immensely powerful but, unlike a one-time pad, could be broken with sufficient effort.

The problem with any key-based encryption system is getting your key securely to the recipient without anyone else gaining access to it. As soon as the key is intercepted, there is no secret. That's fine if I can visit you and give you the key physically beforehand, and we can both keep our copies of the key securely locked away. But that's a labor-intensive process, and simply not possible in many circumstances, whether reaching a distant battlefield or in a more mundane modern application like keeping your credit card number secure when you enter it online to buy from a Web site. The answer (without equipping everyone with an Enigma machine) came in the form of public key encryption, the technology behind pretty well all online security. It works with not one key but two.

The person who is going to receive the information gives out a public key to anyone who wants it. This key is a magic number that is used to encrypt the information. But the information can only be decoded with a second key—and *that* key is only known to the receiver. Because of this, it doesn't matter if someone eavesdrops and picks up the key the sender uses to encrypt the information, because that public key isn't enough to decode the message. In chapter 6 we'll come back to how such a public key system can be made to work—and how entanglement may be its downfall.

Public key systems are immensely convenient. And can be very difficult to crack. But unlike the one-time pad, it's always possible, in theory, to break them. Even with the development of very sophisticated public key-encryption systems, scientists and mathematicians continued to look out for an unbreakable technique that wouldn't involve the clumsy distribution of a one-time pad key. What was needed was a mechanism for distributing information that would enable encryption and decryption, but whose key couldn't be intercepted by an eavesdropper. And the surprise mechanism that looks likely to provide such cast-iron security is quantum entanglement.

The scientist who first came up with a practical scheme to use entanglement to keep secrets was Artur Ekert. In 1991, Ekert described a way that an encryption key could be sent from one place to another with absolute certainty that it had not been intercepted. Of Polish/British extract, Ekert had a colorful early life traveling in the United States and Europe. He took his first degree at Krakow University and was involved in the Solidarity movement during that city's turbulent period of Poland's past.

Ekert has worked in the UK since 1991, originally at Oxford and most recently at the University of Cambridge, where he is the Leigh Trapnell professor of Quantum Physics and a fellow of King's College (he is also Temasek professor at the National University of Singapore). Ekert was reading the original EPR paper on entanglement and noticed that Einstein and his coauthors described a process that sounded just like perfect eavesdropping in cryptography. If you can learn about a property without disturbing it, you can intercept a message without being detected. Ekert realized that John Bell's test for entanglement could be used to detect the interception of a message.

By coincidence, John Bell gave a lecture at Oxford soon after Artur Ekert's eureka moment. When Ekert told him of his idea, Ekert reports that Bell said, "Well, gosh, I didn't think it would ever be practical." He was very surprised at the suggestion of the Bell inequalities being put to work.

Ekert's method involves having a central source of entangled photons. One photon of each pair is sent to each end of the communication link. The two recipients measure the polarizations of the photons at one of three angles, randomly selected on each measurement. They then

Figure 4.2. Entanglement-based quantum encryption.

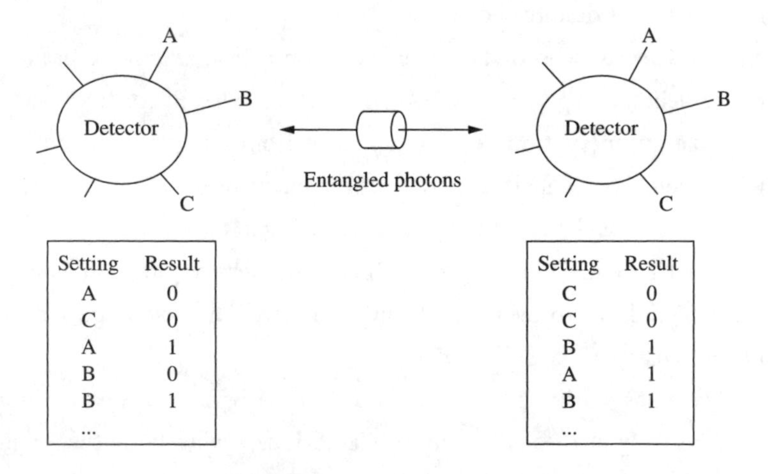

share some information that will be of use to them, but does not have to be kept secret: which angles they used for each measurement, and the results of measurement when they used *different* angles from each other.

In the example above, before each measurement, each detector is set randomly in one of the directions A, B, or C. When an entangled photon reaches the detector, it will come up with a result of either 0 or 1 when measured in that direction. After making a series of readings, the two ends of the link share the position the detector was in (A, C, A, B, B . . . for the left-hand detector and C, C, B, A, B . . . for the one on the right). Where the directions were *different,* they also share the result. So the values shared from left-hand detector are 0, –, 1, 0, – . . . and the right-hand 0, –, 1, 1, – . . .

These readings are checked for how frequently they are the same—if the photons were still entangled, and so not tampered with, there will be closer correlation than is expected without the spooky connection of entanglement. It's possible to tell from the compared results that the photons are still entangled. And that means they hadn't been intercepted, because any attempt to eavesdrop and read the values would break the entanglement.

So the photons that *were* measured at the same angle (in the example the second and fifth photons) are safe and secure. We know that they weren't eavesdropped on, and there has been no information broadcast about their values. And because of the entanglement, we know that they both had the same value, a value known only to the owners of the two detectors.

You might be wondering now what the point is, because no information has gone from sender to recipient. When does the key get sent? The answer is not only very clever but overcomes one of the biggest problems of conventional private key cryptography. Normally, if I want to send you a message, I first generate a key, then send it to you, then we communicate using the key. But that means there are three opportunities to break into our system. A spy could find the key at my end before I use it, or could intercept the key as it is sent (by whatever means) from me to you, or could find the key wherever you are storing it.

It's often easier to get a look at the key at either the sender or receiver's end than it is to intercept the key as it is transferred. Magically, Ekert's entanglement technique provides the key itself—I don't need to make one up, and neither of us need to store it beforehand. The icing on the cake is that, unlike anything I can generate with a computer, this entanglement generated key will be truly random. (If you think computers can generate real random numbers, see page 161.)

When the two ends of the link both measure polarization on the same orientation, the result with be either 0 or 1, totally at random. We both get the same result, but we have no idea what that result is going to be. And that random sequence of 0s and 1s becomes the key itself.

Even cleverer, if there were some way to store the entangled photons, we could keep the photons that will become the key as long as we like, and it is only at the point that we make the measurements that the key itself comes into being—so no one can sneak in and take a look at the key beforehand. In practice, storing photons is not easy, though several techniques are now emerging (see page 196 for instance, for a method of entirely stopping light), but in practice it probably won't be necessary to keep the entangled photons on the shelf, they can be picked up when required—perhaps from a satellite.

This is why the experiments mentioned at the end of the previous chapter are so important. If it is possible for a satellite to be the source of "raw" entangled photons, they could be beamed to two locations on the earth and provide a superb long-range entanglement-based source of secrecy.

Artur Ekert's weren't the first attempts to use the strangeness of the quantum world to keep data safe. As far back as 1970, Stephen Wiesner, of Columbia University, had suggested that the peculiarities of quantum mechanics would make it possible to construct paper currency that could not be forged. Wiesner imagined a bill with both a conventional serial number and a set of twenty photons stored on it, each of the photons randomly polarized in one of four directions.

To make Wiesner's curious currency work, banks would have a secret book that linked the serial number to the polarizations. Polarization is a quantum state, making it respond to measurement in a strange way (see page 58)—only if you know in advance what the polarization of each photon is can you measure it with certainty. Think about that for a moment—you can only discover a photon's polarization with total accuracy if you already know what the result should be.

For instance, if you have a photon polarized at 45 degrees to the horizontal and check if it polarized horizontally, the answer won't be "no" or "1/2"—instead you will get "yes" 50 percent of the time and "no" 50 percent of the time. All that taking a measurement horizontally and getting "yes" tells you is that the photon wasn't polarized vertically—but it could have had any other orientation. Similarly, if you get the result "no," all you know is that it isn't horizontal, but not which of the other polarizations it has.

However, if you had a copy of the bank's secret book, and you knew the polarization should be at 45 degrees to the horizontal, and measured it that way, it would come up as "yes" every time. Because it's not possible to read the values of the polarizations of the photons, you can't duplicate them. So the only currency with a particular serial number *and* the whole set of twenty photons with the right polarization as specified in the bank's secret book, would be the real, original bill. The neat trick that a known polarization can be checked but an unknown polarization can't be measured, makes it impossible to reproduce the values from the banknote.

Of course, Wiesner's idea, though highly entertaining, is both incredible and impractical. Back in the low-technology world of 1970, what he suggested was impossible to achieve in any circumstances, and even now the technology to keep twenty photons suspended in limbo would cost hundreds of thousands of dollars and would be the size (and weight) of a small oven—hardly an effective way to protect a dollar bill from forgery (though it might be worth it if there were some way to embed such a forgery detector into the *Mona Lisa,* say). This impracticality was reflected in the reaction Wiesner got at the time. He ruefully remarked:

> *I didn't get any support from my thesis advisor—he showed no interest in it at all. I showed it to several other people, and they all pulled a strange face, and went straight back to what they were already doing.*

Even if you could make money that contained twenty polarized photons, the sheer act of validating a bill destroys it, which misses the whole point of being able to check for forgery. Yes, banks could reinstate the value, but stores couldn't. And that's after swallowing a very big "even if." How could you keep a set of polarized photons in a bill? Photons have a habit of traveling at the speed of light—they don't sit around and wait to be measured. Wiener imagined each photon sitting in a tiny, perfectly made mirror-lined box, bouncing back and forth for eternity. Nice in principle, simply not achievable in reality.

However, some of the greatest breakthroughs in science have originated as impossible thought experiments, and it's likely that Wiesner's brave folly sowed a seed for the next ideas on using quantum elements in cryptography. These ideas would come from Charles Bennett, who had been a friend of Wiesner's when an undergraduate. Bennett, now a scientist working for the computer manufacturer IBM, and Gilles Brassard, of the University of Montreal, developed a much more practical way to use quantum theory to keep something secure.

Born in 1943, Bennett started his academic life as a chemist but, from 1972, when he went to work for IBM, he majored in the crossover between physics and information theory. Although best known as a commercial enterprise, IBM has always invested strongly in pure research, and Bennett's work would demonstrate the benefits of that strategy. Bennett's collaborator, Gilles Brassard, was born in Montreal in 1955, and had only recently become a faculty member at his university, his first academic job, when they began to work on quantum encryption.

Bennett and Brassard devised a scheme for distributing a secret key across open channels. As we have already seen, this is a dream for cryptographers, as private keys are the ideal way to keep a secret, but there is a huge difficulty is getting the key safely to the recipient without it being intercepted. Bennett and Brassard imagined using a stream of photons as the transmission method of the key, with the direction of photons' polarizations providing the 0 or 1 bit (for example, if they measured horizontally, a horizontal polarization result could be 1 and vertical 0).

The cunning part is that both transmitter and recipient randomly vary their measurement between horizontal/vertical and diagonal. After the key has been transmitted, each end of the transmission tells the other which orientations they used (but not whether they carried "0" or "1"). This means they can discard the measurements taken when the orientation at one end was different from the other.

As with Ekert's later entanglement-based scheme, the clever bit is that, if a third party had intercepted the photons along the way, they would have disrupted the measurements. At the time the intervention took place, no one had shared what orientation of detectors was being used, so the eavesdropper wouldn't know what the right settings were. The act of measuring would change some of the polarizations. If the two communicators then shared the value of some of their photons (and discarded those photons, not using them to send the key), they could check if there had been eavesdropping.

Only when the two ends had agreed on enough random settings of their devices to cover not only the key itself but their throwaway check photons would the key transfer be complete—and only if the checks came up clean would the key be used. It seemed infallible. But there was significant resistance to the idea. This time it wasn't the theory that was doubted—unlike Wiesner's, Bennett and Brassard's idea was widely supported—but the possibility of ever doing it in practice was considered unlikely. Quantum theory was just that—a theory. It seemed too insubstantial to be the basis of a communication system.

In the end, Bennett felt that the only way to win over the doubters was to move from theory to experiment. More a theoretician than an experimenter himself, he roped in a graduate student, John Smolin, to help him construct the quantum encryption equipment. It was a learning experience for both of them. As Smolin later commented:

> *Neither Charlie nor I knew much about building anything, but we knew enough to be dangerous. As an example of Charlie's experimental agility, I remember a time I was visiting George [Dyer—Bennett's stepson] in their apartment in Cambridge, Massachusetts. Charlie was excited about some fancy new tea he had gotten somewhere. He had set up a little double boiler using a pot and a teapot, explaining how this was the right way to cook the very delicate tea. George and I left the house for some time,*

returning hours later; when we came into the kitchen, we noticed the teapot. If you know about blackbody radiation, you've probably seen how a blackbody turns invisible in a furnace, radiating the same spectrum as fills the cavity. The situation we found was not quite that, but there was a red teapot sitting in an empty pot on the stove. This would not have been disturbing, except for the fact that at room temperature the teapot had been green. I had forgotten this and wouldn't have known except that George pointed it out, and proved it by turning off the stove. The delicate tea had left nothing but a faint burnt aroma.

Apart from any practical limitations the pair had as experimenters, there was also the minor matter of having no budget with which to buy equipment. Theoreticians like Bennett usually needed little more than pencil and paper—building an experimental rig was a different world. Parts were dug out of IBM's storeroom, and Bennett discovered he could bend the bureaucracy, ordering almost anything as long it cost less than $300 and so wouldn't be counted as a capital expenditure.

In 1988, at around three in the morning in a pitch-black room, Bennett and Smolin managed to get a positive result. Two computers, admittedly only 1 foot apart, successfully exchanged a key with apparent security thanks to the quantum method (the experiment didn't at this stage test the effect of eavesdropping). The light-tight box linking the two computers, used to keep the encrypted photons from stray interference, was dryly if rather oddly referred to as "Aunt Martha's coffin." A passerby had given Smolin his condolences after noticing the box, draped in black velvet, on Bennett's desk. The velvet was there to plug a light leak, but it gave the box a darkly funereal air.

Bennett and Brassard's idea had been proved workable in the lab, and is already the basis for some early ventures into commercial quantum cryptography, but their approach has a number of problems if it is to be developed into a widely usable product that can be relied on in the real world. First, it only works if you can send individual photons from place to place, something that was impossible when Bennett and Brassard were working on the idea in the 1980s, and still remains difficult today in a robust commercial product. This meant Bennett and Brassard had to use a burst of polarized photons each time, rather than a single photon—and in principle an eavesdropper might be able to steal a few photons from each burst without being detected. This approach also requires multiple communications between the two ends to send the photons, compare settings, compare checks, and decide whether to commit with the key.

And that's not all. This approach of using polarized photons directly as the quantum key has problems because of the inherent losses and failures that all real devices have, and that are all too often forgotten in hypothetical concepts. The system relies on being able to detect a photon when it arrives at the receiving station. But there will be errors in this process, and they can only be resolved by a degree of communication between the two ends. This system of checks keeps the data robust but opens up the possibility that some of the information will leak out. Bennett and Brassard came up with a technique to avoid this, called privacy amplification, which in principle enables them to clean up their key, dropping parts that might have been overheard.

This technique is itself susceptible to interception if the eavesdropper can use quantum entanglement to extract information more subtly, reducing the impact of the intercept. Even with privacy amplification, direct quantum key distribution like Bennett and Brassard's can always fall foul of interception by entanglement; Artur Ekert's entangled quantum encryption, using an equivalent of privacy amplification that only works with entangled photons, isn't.

Finally, Bennett and Brassard's technique depends on a random key being generated at one end of the process before sending it to the other. Unlike the entanglement version of quantum encryption, where the key is generated by the mechanism itself, this leaves the key open to detection, and at risk of not being truly random unless it too is generated by a quantum process. Entanglement is a little harder to deal with than Bennett and Brassard's approach, but has clear advantages from a security viewpoint.

Using entanglement to generate the key is not the only way that it can help with security. If a conventional private key were used, as in the Bennett and Brassard method, it is also possible to use entanglement to help detect interception. Imagine generating a stream of entangled photons, which are injected into the data stream. They could be inserted at random positions in the data, with only the percentage of entangled photons known.

At the receiver end, the photons could be tested for entanglement. If the percentage was still the same, the key could be used. If it had dropped significantly, then the message could have been intercepted—because reading the message would destroy the entanglement. Of course, such a technique would have to be refined to stop, for instance, the eavesdropper from setting up new entanglements to the right percentage, but in principle the action of sprinkling a key with entanglement could be enough to keep it secure.

Quantum entanglement, used in the way that Ekert suggested, remains the closest anyone has come to devising a means of practical yet perfectly secure communication. But the people who delight in breaking through security don't keep still and the very claim that quantum encryption is unbreakable is enough to get hackers' hackles rising. It ought to be stressed that the people who try to break encryption aren't always the bad guys—they could be working for a national security agency trying to prevent terrorism, or for an encryption developer wanting to test the effectiveness of their technique.

So have the hackers got a chance? Can it be possible to break the unbreakable, to get through an impenetrable barrier? The simple answer is maybe. Remember that a one-time pad, generating a random private key the size of the message, is totally unbreakable, and that has been the case since 1918—but the approach still isn't used very often. Even though the encryption itself may be flawless, the generation, storage, distribution, and use of the pads are all open to security breaches. And in the same way, quantum entanglement's unbreakable monitoring of eavesdropping might be possible to circumvent.

First, there is a very small possibility that there's a flaw in the physics, some further quantum weirdness that will allow for an impossible eavesdrop, just as many would have argued until recently that entanglement itself was impossible to make use of in the real world. If that happens, the breakthrough is unlikely to come from the hackers—it is going to be a conventional scientific discovery that then opens the possibility of breaking the security, just as the development of a quantum computer (see chapter 6) could lead to the easy cracking of the public key encryption used today on the Internet.

More likely, though, is for a gap to be discovered between the physics and the engineering. Although several organizations now have early versions of turnkey quantum encryption systems that can be bought and installed for real use in the commercial world, there is a yawning chasm between a lab experiment and a robust, practical device. And the devil is in the detail of moving from theory and controlled conditions to reality and the dangerous world outside the lab. It's true that no one can break quantum encryption by simply eavesdropping on the line—but there are other possibilities.

One of the dangers of using powerful technology is that people get lax, because they trust the technology to keep them safe. As automobiles have got safer, for instance, drivers take more risks. If a computer's communications are truly secure, the people using it can get lazy and good hackers will be ready to exploit the loopholes that occur not in the science but in the people. Often security that is technically excellent is let down by human laziness or incompetence. People write passwords on sticky notes that they fix to a computer screen, leave passwords at their factory settings, or use passwords like "password," resulting in systems that are wide open to hacking. Users of an entanglement encrypted system could be just as careless.

If you follow a secure message from its creation, down the line through a quantum entanglement system and out the other end, that message has to start off in a readable form, and it has to be rendered into readable form at the far end. At some point, if it has any value, the information has to get to a human brain, and that implies passing through print on a screen or on a page that is free of any encryption. Obviously there's nothing quantum entanglement (or any other encryption system) can do to protect the information once it has been decrypted. So one apparently obvious but easily overlooked opportunity is to break in before the encryption is applied or after it is removed.

This can be done a number of ways, from the trivial—simply watching the screen or the keyboard of the person typing or reading the information—to more subtle means that involve monitoring the signal in the wires between a computer keyboard and the PC itself, or the cable connecting a computer to its screen. Already anyone can buy a simple, commercially available product looking like a harmless connector that can be inserted between keyboard and PC and then captures every keystroke as it is typed.

Frighteningly, there are still very senior people in business, government, and the military who believe reading information from a computer is beneath them and so insist that messages are printed on paper. Here again, there is an opportunity to sneak a glimpse of the message if it ever makes the printed page.

But even if there is no way to intercept the information outside its encrypted form, it may be possible to fool the system itself. Here's an example that wouldn't work, because engineers are already aware of it and so will watch out for it, but something similar might still be possible: real-life systems have to cope with noise and loss on the line. No actual system manages to get 100 percent of the data through first time. That means all communication systems have to be able to cope with errors, either by using redundancy—repeating the elements that make up the message—or by having mechanisms for detecting errors and asking for the information to be sent again.

Most quantum encryption systems rely on establishing a series of 0 or 1 values that will become the key that is used to encrypt the real message. Imagine that a hacker managed to send a powerful pulse down the fiber optic linking the two ends of the communication channel, a pulse that was so strong that it burned out one of the two detectors at the receiving end. If one of the detectors could be put out of action, the "random" pattern that was built up would consist of all 1s, say. The key would be a string of 1s. Yet the system's operator would assume that the encryption was still making use of a key that varied unpredictably.

Equally, it may be that some part of the system will unwittingly give out information. If an eavesdropper can break into the circuit and monitor not the traffic itself—which would upset the quantum measurements and result in detection—but something less obvious, like the control signals to a laser, or if an eavesdropper could transmit their own photons up toward the sender and get a different response from the equipment depending on what settings are being used at that moment, it might be possible to tap into the system without upsetting the entangled link.

None of these are definite possibilities. Entanglement remains the most powerful vehicle imaginable for encryption—it's just that, when turning a scientific theory into a practical engineering solution, there are all kinds of possibilities that have to be explored.

Quantum cryptography is very close to wide-scale commercial acceptability, and is likely to be used significantly for important secure messaging by the year 2010. Not surprisingly, there is considerable interest from both security organizations and financial institutions. On April 21, 2004, Anton Zeilinger, the showman of the quantum world, raised the stakes for future customers of such a system by demonstrating a link between Vienna City Hall and the Bank of Austria. We have already met Zeilinger several times and he has had a crucial role to play in opening up the practical side of entanglement.

Anton Zeilinger, born in Austria in 1945, became interested in the peculiarities of quantum physics by accident when reviewing for his finals at the University of Vienna. "I didn't go to a single quantum mechanics class. I learned it all at the last minute before the final examination from books. What I read excited me—perhaps more so than otherwise." Ever since, experimenting with the quantum world has been an obsession for Zeilinger.

The communication link that Zeilinger and his team set up from City Hall to the branch of the Bank of Austria in the nearby street called Schottengasse was secured with entanglement-based quantum cryptography, and using off-the-shelf software (except for the encryption and decryption modules) managed to transfer 3,000 euros (around $4,000 at the time) from the mayor of Vienna's funds into the university's account, a gift from the administration toward their research.

For this demonstration, the encrypted link was set up over a distance of around 500 meters (about one-third mile), with optical cables threaded through Vienna's ancient sewers, previously best known as one of the settings of the classic Orson Welles movie *The Third Man.* But, of course, Zeilinger had already demonstrated that it's possible to send entangled photons long distances through the air, and, in 2004, the Singapore government decided to combine this sort of free air network and quantum encryption.

Singapore plans to be the first state with a nationwide communication network protected by quantum encryption. To be fair, this is a lot easier for Singapore than most countries.

The main island of Singapore is only around 15 miles by 8. By broadcasting entangled photons from the center of the island, all important locations could be covered by a maximum spread of distances similar to those that have already proved practical for entanglement broadcasts in Vienna.

The project, a joint effort between the Temasek Laboratories defense establishment, the National University of Singapore, Nanyang Technological University, and the government-funded A*STAR agency, began by building links over about a kilometer (two-thirds mile)—where the Vienna team only had to keep the link going long enough to run the experiment, the Singapore setup will have to keep underway all the time despite the worst the weather can do, which implies a lot more testing to see how the technology stands up to wind, rain, air currents, and even earthquakes.

Like the Austrians, the Singapore team, made up from academic, business, and government staff, will need to employ active optics to keep the beams on track, but the intention is to have a star-shaped network crossing the island, available to both business and defense interests, before 2010. With significant commercial activity in the United

States and Europe on both fiber-optic and free air quantum encryption systems, there is little doubt of worldwide adoption.

Entanglement had proved its worth in keeping messages secure, and is likely to have its first commercial application in this field—but this isn't the only aspect of communication that has been linked with entanglement. After all, the spooky connection between entangled particles seems to defy relativity in the way that it responds instantaneously at any distance. Is there an opportunity to break the light barrier and send an instant message through this nonlocalized link?

CHAPTER FIVE

THE BLISH EFFECT

My interest was fanned by two coincidences—the kind of coincidences that cause and effect just can't allow, but which do seem to happen all the same in the world of unchangeable events.

—JAMES BLISH, *The Quincunx of Time*

In 1954, science fiction author James Blish wrote the short story "Beep" to explore the consequences of instantaneous communication across any distance. The ability to send messages anywhere, instantly, has a powerful appeal, and it's a possibility that occurs to almost everyone who hears about quantum entanglement.

After all, instant communication is exactly what the entangled particles are doing. A change in one is reflected in the other, wherever they are, at whatever the distance. If you could harness that uncanny power, if you could use entanglement to send instant messages at any range, it would prove immensely valuable, not only as we've already seen to keep the message secure, but also to overcome light lag.

Even on the relatively small scale of the Earth, the speed of light, the fastest rate at which we can send a message (around 300,000 kilometers or 186,000 miles a second in a vacuum, and slower elsewhere), causes delays. Light speed introduces irritating pauses into a satellite phone conversation, and provides practical headaches for engineers building large-scale communication networks. The farther information has to travel, the greater the lag becomes.

A message from a Mars probe such as the rovers Spirit and Opportunity, when Mars and Earth are at their average distance apart, takes four minutes to reach us. This is a lengthy delay that makes any real-time control near impossible. From the nearest star (other than the Sun), any communication would take more than four years. And from the farthest reaches of the galaxy, it would be billions of years before a message could arrive. Not exactly high-speed communication. It was against the backdrop of getting a message across the depths of space that Blish set his story.

James Blish, an American who spent much of his working life in Europe, could put his hand to almost anything that fell within science fiction's broad scope. His most feted book, *A Case of Conscience,* is a thoughtful psychological exploration of what it is to be human; yet he also wrote the Cities in Flight series, a sweeping galactic saga worthy of Isaac Asimov in breadth and imagination, a pile of Star Trek novels, and even a fictional life of the medieval proto-scientist Roger Bacon.

"Beep" was different again. Blish himself admits in the preface to *The Quincunx of Time,* a novel based on an expanded version of the story, that "there is . . . not much physical action here, let alone any melodrama. The structure of the story is still nearly as skeletal, indeed nearly perfunctory, as [my editor] Mr. Sloane held it to be in 1954." This is an almost pure idea story. Blish came up with an extraordinary concept, then used fiction as a vehicle to explore the consequences of his idea.

"Beep" is set in the year 2090. Faster-than-light space travel has been possible for some time, but there is no equivalent for messages. The quickest way to communicate across the light-years is to carry a letter in a spaceship. But a new invention, the Dirac transmitter (named after twentieth-century British physicist Paul A. M. Dirac), is about to change everything. With the Dirac transmitter, communication is instantaneous, whatever the distance. And the availability of instant communication has a huge and unexpected impact on everyone's lives.

Blish does not say whether his story was a conscious reflection of history, though his book on Roger Bacon made it clear that he was fascinated by the historical context of science. Whatever "Beep"'s influences, there is such an important parallel in the development of modern communication that we need to ripple briefly back in time to the 1800s to follow the transition of a message from a physical object to a virtual entity.

Since animals were first domesticated, the fastest mechanism for transmitting a complex message across land had been the horseback messenger, whether carrying a verbal communication in his head or a written document in his saddlebag. Where water intervened, the transmission rate for information came down to the speed of a sailing boat. A message from just a few hundred miles away could take days—if the distances involved were continental, the lag would be weeks or even months.

The limitations of a horse—galloping at around thirty miles per hour, or over long distances making maybe only ten—wasn't an absolute restriction for the transmission of information, even in ancient times. The important qualifier here is the word "complex."

When the only requirement is a simple alert, church bells can pass a warning at the speed of sound—around 760 miles per hour at sea level. If there is a line of sight between source and destination, a basic signal can be transmitted at the speed of light by anything visible at the required range—displaying a white flag or sending up a smoke column during the day, for example, or lighting beacon fires at night, all techniques that have been widely used across the world from the alert stations on the Great Wall of China to the smoke signals of Native-American tribes.

Such techniques were fine to grab the attention, but were of little use for the sort of message that could convey practical news. To say who had just won the World Series, to give details of a scientific break-through, to declare undying love, undertake a business deal, or declare war meant going back to the steady plod of the horseback rider. It wasn't until the 1790s that Frenchman Claude Chappe realized that, by combining the limited but speedy systems used to raise the alert with an appropriate code, it would be possible to pass on a much more de-tailed message.

Chappe started out by making use of sound—combining a loudly banged gong (actually a massive metal casserole pot, or marmite) with a pair of clocks. Once the clocks had been synchronized by a series of clangs, the marmite would be hit as the second hand of the sender's clock passed a chosen number. The receiver would hear the sound as *his* clock's second hand passed the same number (any lag due to the speed of sound was built into the system by synchronizing from those initial clangs). It then only required a conversion from the clock's numbers to an appropriate code. To make it even simpler, letters were pasted onto the clock face.

Sound has the useful ability to work its way around obstacles but dissipates very quickly and is highly dependent on wind direction. It's dangerous to expect sound to travel reliably for more than about half a mile, which would imply having a huge number of stations to carry a sound message across a country. Sound is also unpleasantly intrusive.

Church bells or a muezzin's call, the long-distance audio communication systems most regularly still in use, may be acceptable to call worshippers to prayer, but would soon become unbearable if they were used to transmit messages constantly, day and night. So Chappe, once satisfied with the basic principle, began to experiment with alternatives.

The obvious way forward was to use sight. After all, visual beacons had been popular since ancient times and could be seen around 10 miles away with good visibility (and perhaps a little help from a telescope). Chappe's first attempt involved simply substituting a visual flash for the audible crash of the marmite by turning over a pivoted wooden panel painted white on one side and black on the other. A brief flash of white was used to indicate the position on the clock face, just as the clang had been. Soon, though, it became obvious that the whole messy business with the clocks could be forgotten if more than one piece of information was transmitted at the same time, enabling any specific letter to be indicated by a combination of "bits" of information.

This could have been achieved using several panels together (and was, in later competing devices), but Chappe chose to use a scaled-up replica of perhaps the most ancient visual form of signaling, waving your arms in the air. His device, mounted on top of a tower, consisted of two massive arms, each with a rotating end piece. By putting the arms and the end pieces in different positions, a string of letters could be passed from station to station. A friend of Chappe's with a classical education suggested he called his long range signaling device a *télégraphe,* from the Greek for "afar" and "one who writes." (Miot de Mélito, the friend in question, comments in his memoirs that Chappe wanted to call his invention a *tachygraphe,* roughly meaning a fast writer, but was told disdainfully by Miot that this name was "bad." That term was later

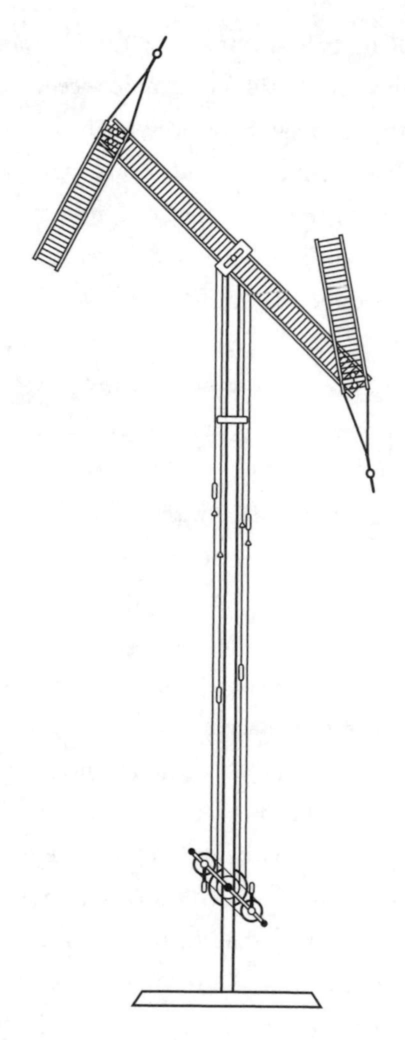

Figure 5.1. Chappe's *télégraphe.*

adopted instead for distance recorders used to monitor the hours that truck drivers spend on the road.)

By 1794, just three years after Chappe converted his system to visual signaling, a line of telegraph stations had been constructed that could send a message 130 miles from Paris to Lille, replacing a long day's ride on horseback with a few minutes of frantic mechanical gesturing between fifteen of the great signal devices. Such was the success of the French telegraph that over the next forty years approximately one thousand telegraph stations would be installed around the world, mostly along lines of strategic defense. This was no accident—the telegraph (or "semaphore" as it became known in English-speaking countries, from "signal bearing" in Greek) was expensive to operate and frustratingly limited by weather and darkness. Business and private messages would have to wait for a cheaper medium that could span greater distance with more reliability.

What that medium should be was obvious even then, if only the practical difficulties could be overcome. It was already known before Chappe's mechanical monsters were constructed that electricity could send a signal down a wire at immense speed—but it was not until the 1830s that parallel developments either side of the Atlantic overcame the two major problems that faced anyone trying to construct an electric telegraph: how to convey information through something invisible like electric pulses, and the more technical issue of getting a signal to travel down a long enough wire to make the system worthwhile.

Electricity had been used in such wild and impractical inventions in the early years of the century that the taint of its past made it difficult for anyone who hoped to base a new invention on it. Both Samuel Morse, who began work on the electric telegraph in New York in 1832, and the team of William Cooke and Charles Wheatstone, who started on the idea independently in Great Britain four years later, were initially labeled cranks.

After years of work and demonstrations that rarely impressed, Morse was finally able to construct a telegraph line alongside the railroad track between Washington and Baltimore, and sent the first message, "What hath God wrought," on May 24, 1844, using his code of dots and dashes to turn electrical impulses in the wire into a signal that could be interpreted by a human being.

Meanwhile, across the Atlantic, Cooke and Wheatstone were to follow a few months later with a device that spelled out letters using the positions of a pair of indicators, pulled around electrically rather like the needle on a magnetic compass. Their electric telegraph line linked London with the otherwise obscure location of Slough, 20 miles to the west. Slough was chosen for its strategic placement as the nearest railroad station to Windsor and Eton, respectively the Royal Family's principal home and the school where the cream of the rich and aristocratic sent their sons.

This new link was to achieve huge publicity by carrying the news of the birth of Queen Victoria's second son from Windsor to the capital on August 6, 1844. Its appearance in *The Times* was said to have reached the London streets within 40 minutes of the news being announced in Windsor, and made much of the telegraph's contribution to a publication rate that even today seems remarkable.

The electric telegraph blossomed with a rapidity that has been realistically compared with the growth of the Internet at the beginning of the twenty-first century. By 1850, there were 12,000 miles of telegraph wire crisscrossing the United States. And when, in 1855, Colonel Arthur Sleigh created a new newspaper in Britain (largely as a vehicle to publish attacks on the Duke of Cambridge, a general in the Crimean War whose promotions of junior officers were based more on social status than ability), it was natural to call his new paper *The Daily Telegraph.* Like *The Times,* it is still in print today.

There was some confusion over just what was going on over the telegraph wires—a man in Maine, for example, watched as an operator tapped an urgent signal down the line, but couldn't understand why the paper slip on which he had written his message was just put on a spike and didn't seem to be going anywhere. When he asked why his message wasn't being sent, and was told it already had been, he assumed he was being lied to because "there it is now on the hook." Even so, thousands of messages were soon flowing. And it was the impact of these messages that enables the telegraph to provide an insight into the impact of instant communication. Let's look at two specific stories, both taken from Tom Standage's excellent book on the development of the telegraph, *The Victorian Internet.*

On January 3, 1854, John Tawell murdered his mistress in Slough, and immediately caught a train to the safe anonymity of London. Only a few years before, the fastest way to send a message from Slough demanding his capture would have been on another train. Back then, by moving so quickly, the murderer could, in effect, manipulate time. Once he had caught the train he traveled back to a point in time where no one around him knew about the incident. As far as London was concerned, he was an innocent man and the murder was yet to happen.

Unfortunately for Tawell, he picked the railroad line with that first British telegraph wire running alongside it. Noting his distinctive long brown coat, the Slough operator telegraphed to the London police to look out for a man "dressed like a kwaker [Quaker]," there being was no *q* on the Cooke and Wheatstone system. Tawell was caught and subsequently hung. The message had undermined his time travel, reaching London long before he did.

Even more striking was the discovery that it was possible to cheat betting systems this way. As soon as anyone in the present day dreams up a scheme for time travel, one of the first applications that springs to mind is discovering the winning lottery number ahead of the draw and making a fortune the easy way. In the nineteenth century, something like this was briefly possible. Back then, bookmakers who took bets on horse racing had a pragmatic attitude to time that reflected the speed of communication available. As far as they were concerned, the race had not ended until news of the result reached the town in which the bookmaker operated, so bets could still be placed on a distant race up to a day after the running had finished.

Canny profiteers quickly spotted that the telegraph enabled them effectively to pass a message back in time, providing the result hours before the bookmaker considered the race to have finished, and allowing for sure-fire betting. This opportunism was not missed by the telegraph companies, whose attempts to stop it led to the early adoption of coded messages by those who wanted to break the law.

In one documented example, a telegraph message was sent around 20 miles from London to Epsom Downs station, close to a famous race course, asking a friend to send some luggage and a shawl up to London. The reply said, "Your luggage and tartan will be safe by the next train." "Tartan" was a code word identifying the colors of the winning jockey, allowing a bet to be placed before the result was known in London.

To a world brought up on letters carried by Pony Express, the electric telegraph turned everything upside down. Compared with anything that had come before it provided instant messaging.

If the telegraph transformed Victorian business and society, so in Blish's imaginary future, the Dirac transmitter was to prove even more revolutionary. In the story "Beep," as the transmitter is about to be tried out for the first time, something remarkable happens. An unknown informant describes the secret test *before* it takes place. Stranger still, in later years the transmitter seems to give the government an uncanny prescience. When an invasion fleet materializes in space, opposing vessels are waiting there, ready to prevent the attack. Yet even if a message had been sent instantaneously as soon as the unfriendly ships appeared, it would have taken days if not weeks to assemble the opposing fleet.

The hero of the story gradually realizes that the Dirac transmitter somehow receives messages from the future. This ability depended on the way the imaginary transmitter worked. Blish was a careful writer and did his best to make use of scientific fact. If he had come across quantum entanglement, he would almost definitely have incorporated that; as it was, he devised a very similar fictional concept. One of his characters, in the sort of speech that demonstrates exactly why Blish's editor felt it necessary to complain, tells us:

> *A positron in motion through a crystal lattice is accompanied by de Broglie waves which are transforms of the waves of an electron in motion somewhere else in the universe. Thus if we control the frequency and path of the positron, we control the placement of the electron . . . somewhere else.*

The positron, the positively charged antimatter version of an electron, first dreamed up as a theory by Dirac and discovered experimentally a few years later, got many scientists (and science fiction writers) excited in the 1950s—Asimov's famous robots, for instance, had "positronic brains"—so it's not surprising that Blish dragged the positron into his imagined technology. That apart, what he describes is unwittingly similar to entanglement. But how did the Dirac transmitter in the story allow its users to know the future?

At the start of each message was a high-pitched beep, an apparently unavoidable piece of technical interference. When this beep was analyzed, every message ever sent using a Dirac transmitter, through all time, was condensed into that single blast of sound. Hence the ability to "see" the future—because warnings of events that were taking place could be picked up before the event happened.

The beep itself is pure fiction, but if it *did* exist, the ability to receive messages from the future has huge and frightening implications. It's not just a matter of being able to send yourself the lottery results and make a fortune—or of being able to warn against alien invasion—the whole chain of causality, the way one event kicks off another, can potentially be destroyed.

Just imagine a simple (if pointless) device: a transmitter that can be switched on and off by a remote control using one of its own transmissions. With the transmitter switched on, we send a message through our time-travel device, back a few seconds in time. That message switches the transmitter off before it can send the message. But if the transmitter was off, it couldn't have sent the message. And if the message wasn't sent, the transmitter would still be on. The outcome is a tangled nightmare, a contradiction that seems to undermine the whole basis of reality.

Why should we worry, though, if Blish's beep is pure fiction? Because *any* message traveling faster than light also travels backward in time. If entanglement did mean we could send messages with no transmission delay, at an infinite speed, we would have the technology to build an informational time machine. It wouldn't allow a person to travel back through the time dimension, but we could send a message into the past. This is an inevitable consequence of part of Einstein's special relativity that he described as "the relativity of simultaneity."

Einstein, himself, in a book describing relativity aimed at a popular audience, introduced the concept by using the picture of a railroad embankment that is struck by lightning in two separate locations. He then made what seems a reasonable request:

> *I make the additional assertion that these two lightning flashes occurred simultaneously. If now I ask you whether there is sense in this statement, you will answer my question with a decided "Yes." But if I now approach you with the request to explain to me the sense of the statement more precisely, you find after some consideration that the answer to this question is not so easy as it appears at first sight.*

How, Einstein asks, can we really *know* that the two lightning strikes were simultaneous? What does simultaneous mean for two events that are spatially separated? We can't be in both places at the same time, nor can we be sure that clocks at the two locations are exactly synchronized (whatever that means). The only way we can check for simultaneity, he suggests, is to have an observer midway between the two strikes with a pair of mirrors so that he can see in both directions at once. If the flashes reach the midpoint at exactly the same time, the events are simultaneous.

So far, so good. But let's make the move from the embankment onto the railroad. Einstein goes on to imagine a very long train, traveling along the track from left to right. We also have an observer with a pair

Figure 5.2. Einstein's flashes on the embankment, arriving simultaneously.

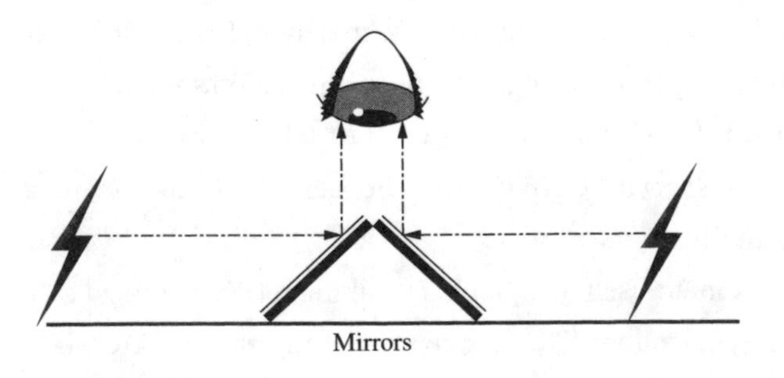

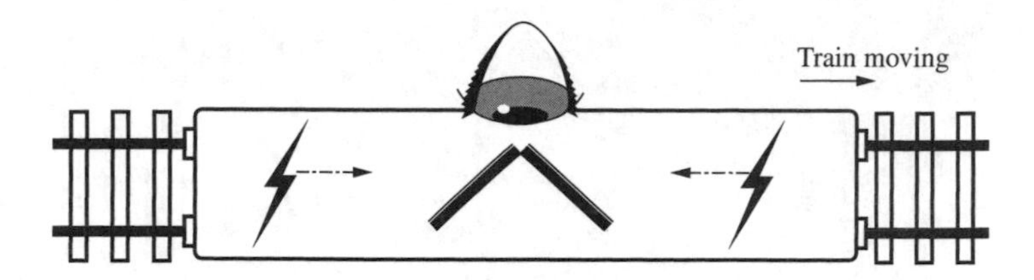

Figure 5.3. Einstein's train at the moment of the flashes.

of mirrors on the train, mirrors that are midway between the lightning strikes at the time the flashes occur.

But now there is a problem. While the light is traveling toward the two mirrors in the center, the train will have moved. If the train is traveling from left to right, then the mirror on the right will be closer to the right-hand lightning strike, while that on the left will be further from the left-hand strike. As far as the observer on the train is concerned, the flash on the right occurs before the flash on the left. Two events that are simultaneous to a fixed observer are no longer simultaneous if an observer is moving. Yet both observers are still perfectly correct.

This might be the best time to get a word in about that concept of a "fixed" observer. In relativity, you always have to put things into context. When I say the observer is fixed, I mean he isn't moving as far as the lightning flashes are concerned. He's sitting still on the ground—something most of us habitually think of as fixed. And that's fine. But we ought to remember that at the same time this "fixed" observer is spinning around with the earth's rotation, flashing around the earth's orbit, and streaking across the universe with the Milky Way galaxy at hundreds of thousands of miles per second—it's hard to get far in relativity without those little words "with respect to" creeping in. In this case, the first observer is fixed with respect to the embankment and the flashes. (The second observer, whom we've thought of as moving, is, however, fixed with respect to the train.)

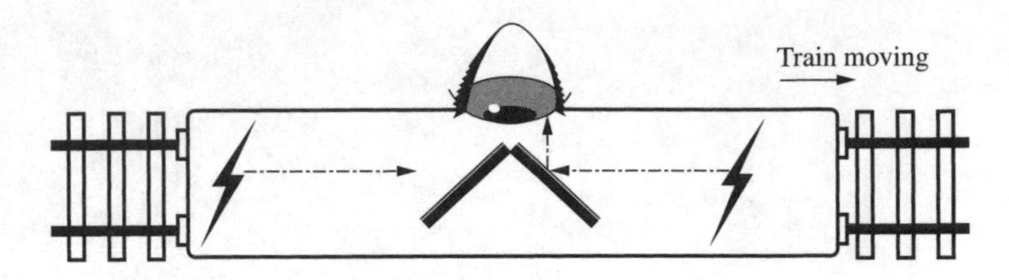

Figure 5.4. Einstein's train a little later: the right-hand flash has arrived, but the left-hand flash is still on its way.

Now anyone who has read a little about relativity tends to get a bit hung up on this train example of Einstein's. At the absolute heart of his special relativity is the idea that light always travels at the same speed, whether you are moving toward the light beam, moving away from the light beam, or are stationary. This fundamental aspect of relativity was responsible for Einstein's conception of it in the first place.

Lying on a grass bank on a park in Bern, Switzerland, Einstein had been letting sunlight play through the lashes of his half-closed eyes. He imagined somehow riding along beside one of the light beams that flickered and flashed in his eyes. If light behaved like everything else, then the light would be stopped, just as a train will be motionless (or would appear to be) if you drive along beside it at its exact speed. The imagined light was not moving at all, as far as Einstein was concerned. And as there is no magic fixed point in the universe, his viewpoint was just as valid as the one that says the light is traveling at 186,000 miles a second. For him, the light was stationary.

This raised a problem. As Einstein knew, Scottish physicist James Clerk Maxwell had shown that light was the interplay of two interconnected natural phenomena. A moving magnet produced electricity. Moving electricity produced magnetism. Light was the ultimate example of pulling yourself up by your own bootstraps. A traveling wave of magnetism supported a traveling wave of electricity that in its turn supported the magnetism—a bit like the party game where a circle of people all sit on one another's laps, each supporting the next. But this could only work if the electromagnetic wave traveled at one particular speed—186,000 miles per second in a vacuum. This speed is such a definitive signature, it is how light was identified as one of these electromagnetic waves in the first place.

Einstein realized that light must behave in a way that seems totally unnatural in comparison to any real-world object that we encounter. However fast or slow we move, light would always travel at the same speed. It was impossible to slow it down by moving away from it, or speed it up by moving toward it. Ordinary relativity, the ideas of relative motion that went all the way back to Galileo, simply didn't apply to light.

This strange behavior of light is why Einstein's argument about the train seems odd when it's first encountered. He appeared to be adding the speed of the train to that of the light from the right-hand flash. In fact he wasn't, because other factors come into play. Einstein's special relativity confuses us because light's constant speed results in uncomfortable changes to the nature of time and space. To understand what is happening, we need to explore the strange world of relativity in a little more depth. The fact remains, though, and has been experimentally verified, that Einstein was correct—on the speeding train, the two events are *not* simultaneous. The right-hand flash occurs before that on the left.

If you aren't convinced, the next few pages will show just why Einstein was right. There's a neat visual trick that can help make clearer what happens to simultaneous events when special relativity is applied. The approach is called a Minkowski diagram, before Russian-German mathematician Hermann Minkowski, who devised them in 1908. These diagrams can become more baffling than the events they describe but, if kept simple, they make it much clearer why things happen the way they do—in this case understanding how "simultaneous" doesn't mean the same thing to someone who is moving.

Minkowski diagrams show space and time, laid out flat on the page. Because this is how it is always done (for no better reason), we measure passage through time as a movement up the vertical axis, while any movement in distance is measured along the horizontal axis. (To keep it simple we just consider one dimension in space—there are, of course, three, but we only need one to see what's happening, and it helps a lot to keep the diagram conveniently flat. To draw a complete Minkowski diagram, representing all the spatial dimensions as well as time, would require four dimensions, which is a trifle inconvenient.)

Let's imagine, because we're going to use this setup to demonstrate instant communication, that we've a transmitter back home and a receiver on a spaceship that is shooting away from the Earth at great speed. We'll look at things from the point of view of the observer on the Earth—with relativity you always need to know whose viewpoint you are taking.

Figure 5.5. A static observer.

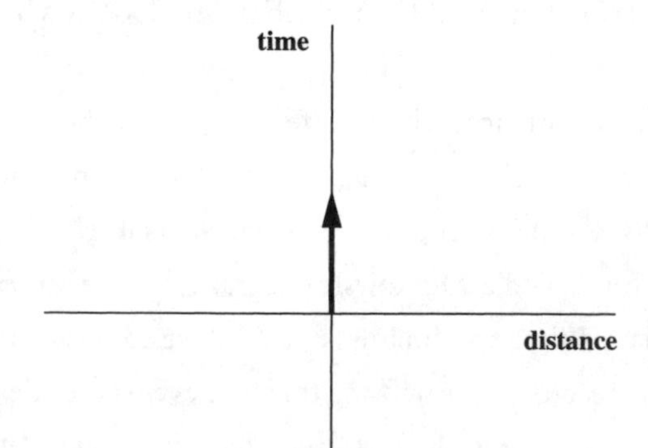

Here's a very simple diagram of the observer, sitting at home without moving (the observer is the dark arrow). As time passes, the observer moves up the time line in the direction of the arrow, because time is ticking on at the rate of a second per second, but there is no side-to-side movement because, as far as the observer is concerned, he isn't moving. An object that isn't moving (according to our observer)—a pen the observer holds, for example—heads straight up the time axis. If the static object were a yard to the observer's right, say, it would be shown as another arrow, parallel to the observer's own, but a yard's distance along the horizontal axis. For convenience (again, this is entirely arbitrary) we start observing at time zero, where the two axes of the diagram cross.

Now let's add the spaceship into the diagram.

The spaceship also flies up the diagram through time at a second per second—but it has a spatial movement, too. It starts right next to our static observer (dangerously close, by the look of it) and heads off into space. To keep things simple, we've ignored messy complications like acceleration. However this ship is powered, it gets up to speed right

Figure 5.6. The spaceship speeds away.

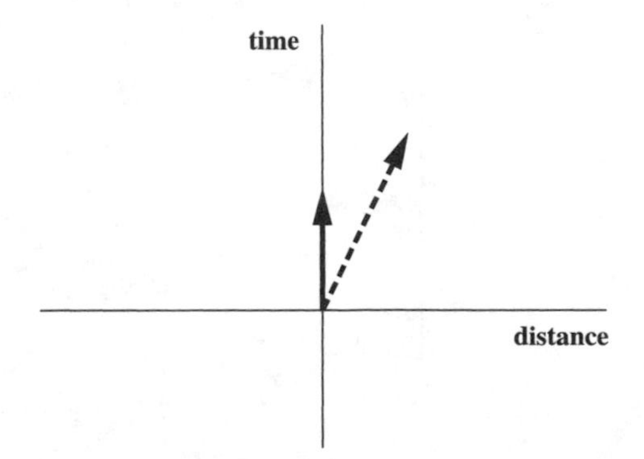

away. As time progresses, it travels away from the observer at a steady rate, so we see a nice straight line. So far, so trivial. But what would a beam of light look like? It would be another straight line, but at a very specific angle.

To keep things really simple, we're going to measure time in years and distance in light-years (i.e., the distance light travels in a year—around 5,870,000,000,000 miles). That means for each year of time that passes, light has traveled one light-year. So the line representing a beam of light would be at 45 degrees, moving exactly the same distance along each axis as it progresses.

The line that represents the spaceship must always be above that of the light beam on the diagram. If it were below, it would be covering more distance in any amount of time than light was—it would be traveling faster than light.

Now, let's think about the passage of time as far as the spaceship pilot is concerned. What does an object that's not moving look like from

Figure 5.7. A light beam.

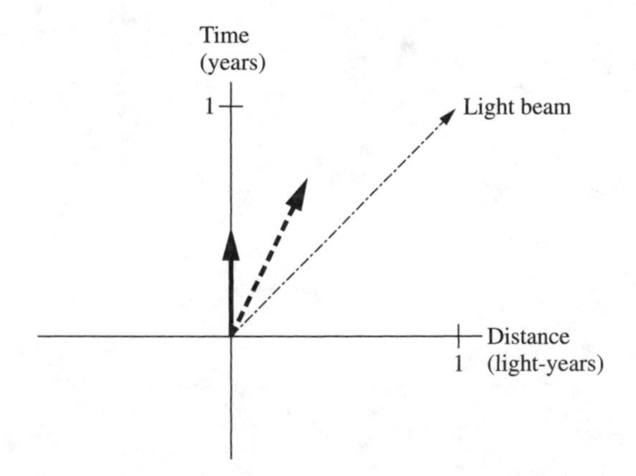

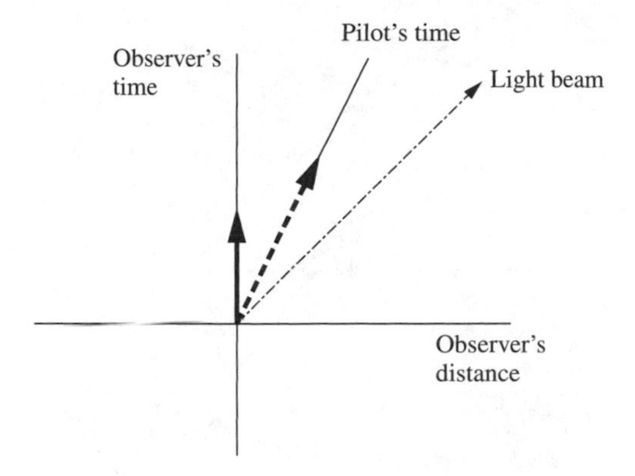

Figure 5.8. The two time axes, from the observer's viewpoint.

her viewpoint? Remember, for the observer on the Earth, an object right next to him that isn't moving must stick with him, moving up the vertical axis as time passes. The vertical time axis, for the observer on the Earth, is the same line as the arrow of a nonmoving object. For the pilot, similarly, an object that doesn't move must stay with her—it has to move up the same line as her. And here again, the time axis *as far as the pilot is concerned,* is the same as the line along which there is no movement from her viewpoint—the line of her movement. Sitting on the Earth with the observer, we plot the pilot's time axis as the line of her movement, quite different from the earthbound observer's own vertical time axis.

Although we are concentrating on time, it's worth reminding ourselves that on the diagram above that the horizontal distance axis is *also* only applicable to the observer's viewpoint—to the pilot, distances also will seem different.

Now, let's cook up a simultaneous event. Einstein's example used

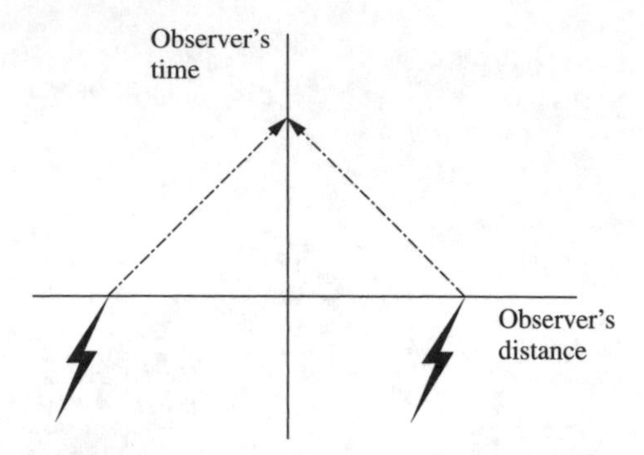

Figure 5.9. Einstein's lightning flashes.

distance to measure simultaneity. His two lightning flashes would look something like above:

The two flashes both start at time zero. Light from the two positions moves in opposite directions but both at 45 degrees (at light speed), eventually meeting at the observer's position at the same time.

We're going to have to use a slightly different version of measuring simultaneous events to apply our spaceship diagram, because Einstein's original example relies on a combination of light speed and equal distances to define simultaneous events. We haven't worked out how distance translates from the earthbound observer to our spaceship, so we need to take a different approach.

Let's call the time where the axes of the diagram cross time zero. Because light always travels at the same speed (at a 45-degree angle on our diagram), I can say that a flash that takes place 10 seconds *before* time zero, bounces off a mirror, and gets back to the same spot 10 seconds *after* time zero, bounced off the mirror at a time that was simultaneous with time zero. In effect, this rotates Einstein's diagram for simultaneous events, so it is symmetrical in time rather than distance.

Because the reflection happened at exactly the same position up the time line as time zero, it was simultaneous with time zero as far as the observer is concerned, even though it is spatially separate from the observer's position.

Let's take a look at a simultaneous event from the space pilot's viewpoint. Here's where special relativity's trickiness comes in. Remember, Einstein's whole development of special relativity was derived from one fact—that light always travels at the same speed, however the person observing it moves. That means that it doesn't matter whose point of view we are looking at on one of these diagrams, light will always travel at a 45-degree angle. It is the one constant while everything else is changing.

Figure 5.10. The observer's lightning flash and mirror.

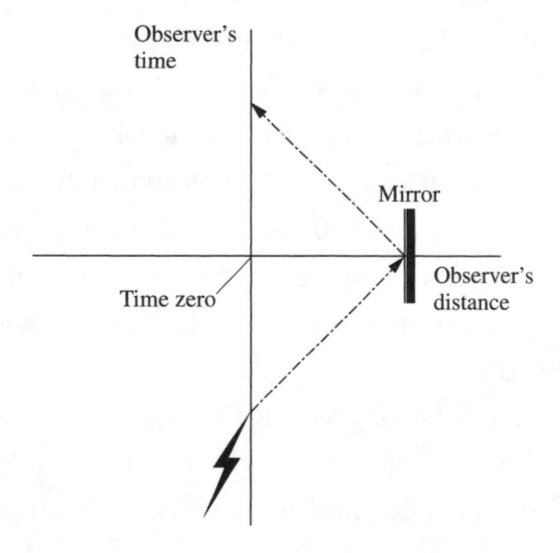

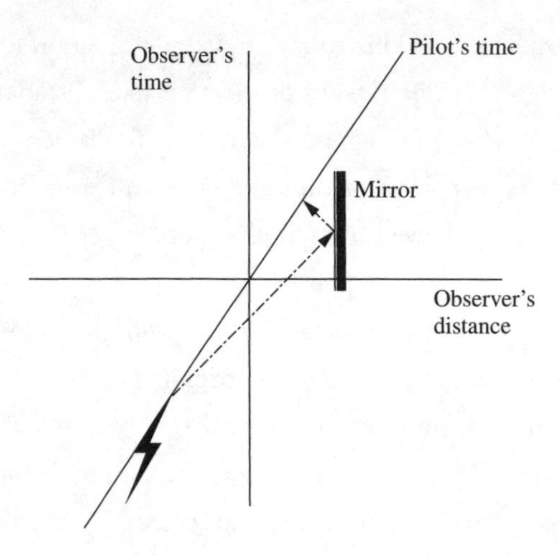

Figure 5.11. The pilot's lightning flash and mirror.

As we did before, we make sure that the flash starts at exactly the same number of seconds *before* time zero as it arrives *after* time zero—but this time, the two points are equal distances either side of time zero on the pilot's time axis, rather than the vertical time axis of the observer. So *from the pilot's viewpoint,* the reflection off the mirror took place at exactly the same time as time zero. It was simultaneous with time zero. But look where the reflection has to take place in order to let the light beams travel at 45 degrees. It's well above time zero on the observer's time line. As far as the observer is concerned, it was shifted later in time. For the observer on the ground, the reflection isn't simultaneous with time zero anymore—just as Einstein's two flashes were no longer simultaneous.

The key to understanding what is happening here is to realize that the passage of time on the spaceship looks different to the observer than it does to the pilot, and that difference comes through in this shift of simultaneity. In the normal world, this shift has little significance because it won't be noticeable, but let's imagine we have sent a probe into space and have accelerated it to a reasonable proportion of the speed of light. The distortion of simultaneity caused by its movement is emphasized by distance. If, for instance, the probe were traveling away from us at half the speed of light and had reached ten light-years distance, then its clock would be running around 5¾ years behind ours, from the Earth observer's point of view.

If I send an instant signal out to the probe, say, in the year 2020, arriving at the same moment *according to me,* it will arrive on the probe when the clock is showing between 2014 and 2015. But this is relativity. From the probe's point of view, *our* clocks are running slow. This symmetry occurs because, from the probe's viewpoint, it's the probe that is static and the Earth is moving away. If the probe responds with its own instant signal, exactly the same effect would occur. It would be received in our year 2009—in total, the message would have slipped back over eleven years in time.

When time travel, or the simpler concept of sending messages through time, is discussed, the whole idea is often dismissed because of a lack of evidence. If it were possible to send messages back in time, surely we would be flooded with messages from the future? As there's a stony silence from tomorrow, the most palatable assumption is that it can't be done. It's an impossibility. (The less palatable reason is that humanity is wiped out before it gets round to sending any messages through time.) Using this technique, though, there's nothing contradictory about the lack of future messages.

Any message-through-time device that depends on Einstein simultaneity has to get a relay station up to high speed and a significant distance from the earth. Those clocks have to have time to build up a difference. In the example we've looked at, the probe would have taken twenty years to reach its position, more years than the time shift that is given to the message. The realities of relativity mean that an instant message can never travel back to a point in time before the launch of the probe that relays it. We have to have already built the technology before we can receive messages from the future.

This is an extreme version of the technique, necessary to produce large-scale time slips. There would be no need to go so far or so fast to send a message back just a few hours to win the lottery—or to jump back a few milliseconds to activate the paradox of the machine that switches itself off before it sent the message that would activate the switch. The amount of the slippage is not the point—just the realization that this technique would make it possible to distort the very fabric of reality. *If* there were a way of sending a truly instant message.

This is so bizarre and unsettling that some physicists impose a "causal ordering postulate" (referred to in a painful pun as the "time COP") that says any two events that are causally connected (one triggers the other) must always come in the same order, however you play around with time. This is a handy test for unlikely concepts, but it is not fixed in concrete—there are a number of mechanisms for time travel that don't break the laws of physics, but are simply impractical (such as utilizing vast rotating cylinders formed out of neutron star matter). The causal ordering postulate doesn't make it impossible to communicate backward through time: it's more like an academic raised eyebrow at anyone who counters it.

So, while instant communication would definitely be a huge boon if we ever go in for serious space travel, it also puts the relationship of cause and effect into danger. Is quantum entanglement the vehicle to provide this remarkable two-edged sword? It's certainly strange enough. And as we have seen, almost everyone who hears about entanglement has the initial thought, "Surely this could be used to send messages faster than light." But we need to dig in a bit more deeply before losing our hold on causal reality.

Whatever entanglement can do, it can't be used until we get one-half of each of a set of entangled pairs physically transported to the other end of the communications link. Entanglement isn't like broadcasting, it's more like a walkie-talkie. A pair of particles that were originally in the same place has to be split up and one sent to each would-be communicator. Only then can the entangled link be used.

The first stage of an entangled communication has to involve a conventional, light speed transmission (or an old-fashioned rocket with a box full of the particles on board). But of itself this doesn't prevent an instantaneous message being sent. It just means that there will be a delay while the link is set up. In the early days of telegraphy, a light speed electric telegraph message couldn't be sent across the Atlantic until more than two thousand miles of cable had been slowly and laboriously laid. Similarly, once both ends of our quantum link have their entangled particles, the light speed delay disappears.

Now we can get down to the realities of signaling through a spooky connection. Let's say we're going to use the very simplest entanglement. It wouldn't be any good for sending real messages, but like Chappe's clanging marmite ringing across the French valleys, it is fine to test the concept.

Imagine we've two entangled particles, one at each end of our communication link and the equivalent of our Morse code dot or dash is going to be the spin of the particles. Spin, as we have seen, is a property of many particles that can have just two values—up or down—when measured in any particular direction. As we saw in chapter 2, thanks to the strangeness of the quantum world, before the spin is checked, the particle is in both states at once. After measuring, it will either become spin up or spin down. Let's assume the entanglement is such that the two particles always assume opposite spins when one is examined. Could we use spin up for dot, and spin down for dash?

Unfortunately, no. Imagine we trigger the entangled link by checking the spin of our transmission station's particle. It's spin up. Immediately, the receiver particle clicks into the spin-down state. There's our instantaneous communication. That second particle has just changed from being in multiple states to the spin-down state the moment we observed our transmitter particle. But we had no way of knowing what the spin of the transmitter particle was going to become before observing it. At the far end of the link, the fact that the receiver particle is spin down tells us that the transmitter is spin up—but that information doesn't carry any message. It was a purely random occurrence.

We shouldn't give up all hope, though. A typical communication has two dimensions—the content (in whatever form) and the time at which the event occurs. Remember Chappe's original form of telegraph that got a message across by combining a single signal (the crash of a marmite or the white flash of his tilting board) with a specific time. Couldn't we apply the same technique with our entangled communication?

Provided we can agree on a way to synchronize time (taking into account any relativistic effects), we could image a communicator consisting of a row of entangled particles. Each second (or millisecond, or whatever), on the second, I examine the next particle down the line. If it's still entangled, I register a 1. If it's not entangled anymore because someone has already looked at the particle at the transmitter end, I register a 0. It doesn't matter what state the particle ended up in—the sheer collapse of the entanglement is enough to signal that something has happened. Once more we've got instant communication in a form rather like one of those old-fashioned ticker tapes that continuously spew out a series of characters.

Unfortunately, this approach also has a fundamental flaw. Discovering whether or not a particle is entangled is anything but trivial. Normally, this requires both particles to be brought together. It is then possible to discover whether they have been interfered with. But this is no use to our would-be instant communicators, separated perhaps by light-years, with one of the two entangled particles at each end of this vast gulf.

There is one last hope, though. As we saw in the previous chapter, one of the options for quantum cryptography is to separate the two entangled particles and send one to each of the communicators. The linked nature of the two particles is used to set up a random set of numbers, which then act as an encryption key. But, this technique is only of any use if the two stations can establish whether or not the particles are still entangled (and hence whether the message is being intercepted)—and this is managed without reuniting the entangled particles. Surely this is a vehicle to enable instantaneous communication at a distance?

Sadly, no. It's true that the two stations can establish whether the signal still consists of entangled particles, but this can only be managed by sending a stream of information as a conventional message between the two stations—a message that has to travel at the (relatively) slow speed of light, crushing any possibility of an instantaneous link.

It is possible to check that the particles are still entangled because of our old friend Bell's theorem (see chapter 2). Entanglement was first proved to exist by comparing the states of particles. Similarly, by comparing information on received particles (in a way that does not break the security of the code key), it is possible to establish whether the entangled particles have been intercepted. But, crucially, it is necessary to send information from one station to another through a conventional, light speed channel to make the comparison. Once again, the entangled link refuses to carry a useful message. Perhaps the time COPS really are out there.

It's frustrating—but that's the real picture of entanglement when it comes to instantaneous communication and messages that travel backward in time. Yes, it's true—if you could send a genuinely instantaneous message (or any message that went faster than light speed), it would be received before it was sent. Yes, it's true—the spooky link of entanglement simply disregards distance and acts immediately. But however hard you try, it is not going to be possible to make use of that link to send a message. It would be like drawing a square circle; it's not something that can be achieved, whatever clever technique you come up with. The only way to get anything usable out of entanglement is to communicate *something* in a conventional way as well, and that takes you back down to light speed at best.

This isn't enough for some observers. They worry that even if a message can't be sent, there is still *something* traveling beyond the speed of light, the something that enables one particle to react to the behavior of another. But this concern really misses what underlies entanglement.

Imagine for a moment we had a block of a very special, highly unnatural substance that is totally rigid. When I push one end, the result is that the other end moves immediately, rather than a wave of pressure traveling down the object. I have just communicated faster than light. The motion took no time to get from one end of the object to the other. (This was very similar to some of the early explanations of light when it was thought to have an infinite speed, before its actual speed was determined.)

The real world doesn't behave like this. Objects as we see them are totally different from the detailed level of individual atoms. Apparently solid objects are in fact mostly open space. When you push one end of a "solid" object like a brick, the motion of the atoms on the face nearest your fingers isn't instantly transferred to the atoms at the far end of the brick. Instead, the first atoms get fractionally closer to the next set. This increases the force on the next set of atoms, an influence that moves at the speed of light, so they then start moving too. This forces the next layer of atoms to move, and so on. Instead of the brick moving as a whole, it ripples—but in a way that is just as indiscernible to the observer as are the gaps between the brick's constituent atoms.

In some ways, we ought to regard entanglement as more like the imaginary totally rigid block. Two entangled particles are not so much two separate entities that are communicating, as a single entity that has spatial separation built into it. It might be more helpful to think of entanglement as something that gets around the concept of space rather than something that communicates across any distance.

Getting around space is something that cosmologists have largely learned to live with. If the currently popular theory of the early history of the universe is true, it went through an incredibly vast inflation, like a balloon suddenly going from almost nothing to something the size of the solar system in almost no time at all. If such inflation occurred, it far exceeded the speed of light. But this is considered acceptable because it was only empty space that was inflating, rather than physical objects moving at this speed: no objects, so no information exceeded light speed. If inflation is allowed such a loophole exit clause from the restrictions of relativity, it seems entirely fair that entanglement should as well.

The good news is, though, that entanglement's power is still of huge value even if it can't be used to transmit a conventional message faster than light. This is already apparent in the way entanglement is being used in quantum cryptography, but will become even more obvious when entanglement is introduced to the number-crunching world of the computer.

CHAPTER SIX

THE UNREAL MACHINE

And now I see with eye serene,
The very pulse of the machine;
A being breathing thoughtful breath,
A traveller betwixt life and death;

—WILLIAM WORDSWORTH,
"Simon Lee"

Communications and computers go hand in hand. It's no surprise that Charles Bennett at IBM was one of the first to suggest a practical use of entanglement. But his idea wasn't to use entanglement to improve communications. Along with other manufacturers and universities, by the 1980s IBM was aware of a huge problem looming in the future: electronic computers were delivering great things—but even though they could be expected to increase in capacity for another thirty years or more, it was already obvious that they would eventually run out of steam. Entanglement would suggest future capabilities for computers that currently we can only dream about.

Computers and the Internet have transformed our lives, arguably more so than any other single invention. Take the activity of writing this book. Less than forty years ago, it would have been a totally different and a much more labor-intensive exercise. The text would have been written out painstakingly by hand, redrafted a couple of times, then passed through that excruciating instrument of literary torture, the typewriter.

If you are old enough to have used typewriters, you might regard them quite fondly, seeing them through the fuzzy haze of nostalgia. There was something physical and real about typewriters. The solid clunk of the carriage return, coupled with the buzz as the platen rotated a few notches, had a comfortably solid ring to it. At least a typewriter didn't get viruses or crash or need expensive new software every couple of years. But that positive glow of nostalgia survives because it's easy to forget the painful process of mistake and correction, throwing away whole pages of effort if there were too many errors, not to mention the correcting fluid, the untidy output that left the paper indented with every letter, and the whole mechanical misery of doing it the hard way.

The typewriter wasn't the only drain on time. To research the material for a book would have meant dedicated months spent in libraries. Now, a lot of research can be done online—particularly with readily available electronic archives of journals and scientific papers—while Web-based library catalogues make it possible to get the books to come to you, rather than your traveling hundreds of miles to them. What's more, projects are under way to digitize millions of books in university libraries, making them available directly online. Libraries are still essential to the writer, but the methods of access are becoming much less time-consuming and much more effective.

Even after finishing a typescript in the good old, precomputer days, there were further hazards to be encountered. Once the text was submitted and edited (provided the only copy had not been lost in the mail), someone would have to re-input all the information, entering it onto a typesetting machine, inevitably introducing new errors into the text. And so it went on.

For that matter, it's not just work that has been transformed by information technology. My groceries from the supermarket are ordered online and delivered to the door, as is my book shopping from Amazon. I can find out what's showing at the local movie theater, check an airline timetable and book a flight and hotel room, then explore maps of my destination, all without leaving my desk. I can have a response to an e-mail query from someone the other side of the world in less time than it would take me to write a letter and mail it. It's all so commonplace that it truly is easy to forget just how much computing and telecommunications have transformed everyday life.

To see how this whole process started, and to appreciate an important parallel for the transformation of computers that quantum entanglement could enable, we need to ripple back to the Victorian age. Before the 1800s, large volumes of regularly collected data simply did not exist. There were one-off exercises, like the Domesday Book, which in 1086 provided a compendium of information on the occupants and wealth of towns and villages of William the Conqueror's new English domain, but these were intensely costly to produce and took years to compile.

Without detailed information, modern government and business could not function. As the demand for facts grew, pressure began to build on those who had to process numbers. Banking was increasing in complexity. Industry was moving away from small cottage operations to bigger establishments with more complex data-processing needs. And governments had developed an appetite for statistics, particularly collections of data about the population, most notably the census, which was originally a Roman idea, but has been taken in the modern sense every tenth year since 1790 in the United States and since 1801 in the United Kingdom.

There was also a growing need to develop detailed tables of numbers—for instance, the logarithms used until the late 1960s for everyday large-scale multiplication, or the data needed for safe navigation—all of which required lengthy calculation. The only method available was paper and pen, wielded by an army of clerks (or "computers" as these clerks were called) who laboriously worked through the calculations. Not only was this approach slow, it was error prone, and the errors would often never be spotted.

It took the remarkable insight of Charles Babbage to realize that there was another way to deal with repeated calculations. Babbage was the son of a goldsmith and banker. Born in 1791, he inherited a fortune from his father and could have spent the rest of his life as a dilettante. Certainly Babbage thrived on the social scene of his time, and much of his work seems to have been undertaken as an excuse to show off his capabilities, as a way to impress the rich and famous at his popular soirees. But Babbage was also intelligent, well educated, and very aware of the practical limitations of manual calculation.

By the 1820s, when Babbage first came up with the idea of automatic computation, the industrial revolution was well under way. Machines were already doing the work of many men and women in manual jobs, whether it was pumping water out of a mine or operating the complex mechanism of a loom. And the loom would figure largely in the introduction of mechanical aids to the world of computation.

Babbage's inspiration is usually traced back to the summer of 1821, when he was helping his friend, astronomer John Herschel (son of the better known William Herschel, who discovered the planet Uranus), check a series of astronomical tables. Going cross-eyed with the effort of laboriously working through a mind-numbing array of figures, Babbage is said to have cried out, "My God, Herschel! How I wish these calculations could be executed by steam!"

Soon after, Babbage designed a machine called "the Difference Engine" that would enable such repetitious calculations to be carried out mechanically. Built around a complex mechanism of gears, a computation was set up on the engine by entering a series of numbers on dials. Once the initial values were established, a crank was turned (the point where the steam power that Babbage had wished for would be employed if engines ever got really large), and eventually the answer was generated, clicking up on another series of dials. Babbage built a part of his design, a working model of a section of the engine, but never completed the whole device, as his mind was already playing with greater, more flexible possibilities.

The Difference Engine was the predecessor of the mechanical calculators that would be used in offices and workshops all the way through to the 1960s. The value of these calculators should not be underestimated. Without them, for example, the atomic bomb project would never have got off the ground. But mechanical calculators were slow and severely limited. They weren't computers in the modern sense—and neither was the Difference Engine. There was no concept of a program to control what the machine did separate from the hardware that undertook the action; everything was "hard coded" into the physical structure of the device. And data had to be directly entered onto one set of cogwheels and read off others. This wasn't a universal machine with the flexibility of a true computer.

Babbage was to be distracted from the Difference Engine by something more, something better. This proved an immense frustration for the British Government, which had sunk over £17,000 (around $1.5M by current values) into the development of the Difference Engine. Such a mechanical calculator would have been of great value to a nation that at the time could truly be said to rule the waves—and hence depended on accurate navigation—and that was faced with a burgeoning need for data processing to cope with its growing worldwide empire. Britain wanted the Difference Engine, but Babbage put it to one side like a discarded toy. It was then a mere working model with around a seventh of the final project completed. His new vision for the future of computation was inspired by a humble weaver's loom.

Weaving patterns into a cloth—particularly using the expensive, ultrafine thread of silk—had been a painfully slow process. It was not unusual for two operators to take a whole day to weave a single inch of material. The French, masters of silk weaving in Europe, were the first to spot the possibility of automating this process. The earliest idea came from Jacques de Vaucanson, a government factory inspector, in the 1740s. He envisaged controlling the threads that fed into the loom with a mechanism not unlike a huge musical box. Inside the controller, a metal cylinder slowly turned as the fabric was produced. The cylinder had metal protrusions along the length of its curved side that moved the controls for each thread, just as the protrusions on a musical box pluck the metal chimes.

This was a significant step forward compared with the manually controlled loom, but it still had its limitations. Each pattern would require an expensive cylinder to be built, and the pattern could only run for as long as the circumference of the cylinder would allow—after than it would be back to the start and would begin to repeat. By 1804, before de Vaucanson's mechanically controlled loom could catch on, the idea was rendered totally obsolete by the revolutionary design of the layabout son of a master weaver, Joseph-Marie Jacquard.

Up to this point in his life, Jacquard was widely regarded as a wastrel who seemed to have little interest in either weaving or mechanical contraptions, but his new loom transformed both the industry and his fortune. Jacquard's idea was to bring to the weaving business the same sort of flexibility that the use of materials like parchment and paper brought to writing.

If your only writing medium is stone tablets, into which you carve each letter, you are limited both in the scale of your output and the uses you can make of it. Paper can, in principle, go on forever and can be used anywhere. Rather than control the loom with a fixed cylinder as de Vaucanson proposed, Jacquard had the inspired idea of programming the pattern with a series of pieces of card, each pierced with holes determining whether a particular thread would be incorporated into the weave. The cards were linked together in a train (each card was joined to the previous one with a piece of material in the original design) so that card after card could build up a pattern of indefinite length.

Jacquard wasn't the first to consider using cards—an otherwise unknown weaver by the name of Falcon had thought of noting patterns on cards back in 1728. But Falcon's cards did not truly control the loom. They had to be held up manually against the loom mechanism—they were more a pattern for hand weaving than an automatic control. But Jacquard's idea was to transform the weaving business. With the Jacquard loom, instead of an inch a day, two feet of material could be produced, a 2,300 percent improvement in productivity. And the pattern produced had the potential to be much more complex than anything that had ever been attempted by hand.

Just how complex a pattern could be achieved was demonstrated by one of Babbage's favorite possessions, frequently displayed at his social evenings: what appeared to be an etching of Jacquard at work but which was in fact a picture woven in silk containing 24,000 rows of thread. The result was a remarkably sophisticated portrait. And Babbage was not only impressed with the weaving capabilities of the Jacquard loom. He made the imaginative leap of realizing that this same mechanism could be used to introduce both data and a program to a new and incredibly sophisticated calculating engine, a conceptual device he called the Analytical Engine.

Babbage never got far with construction of the Analytical Engine—it would probably have been impractical to build with the mechanical precision of the day—but he did put a lot of effort into the conceptual design of what was to all intents and purposes a mechanical computer, even developing concepts like the separate processor and memory that feature in every modern computer. Had he been less of a man of his time, he could have been aided even more in this process by a young woman who has often been described as the first computer programmer.

Ada Byron, daughter of the great Romantic poet, was twenty-seven in 1843 when she published a translation of a paper on Babbage's work written (in French) by Italian scientist Luigi Federico Menabrea. Ada was already a long-term fan of Babbage and his work. She was just seventeen when she first met Babbage, and had been instantly captivated by his enthusiasms. There was even some talk of a possible marriage between the two, but Ada's mother was determined to have an aristocratic match, so in 1835 Ada was married off to Lord William King, who became Earl of Lovelace in 1838, formally making Ada Augusta Ada King, Countess of Lovelace.

The fascination that Ada King (or Ada Lovelace as she is often mistakenly called) has aroused in modern times has led to the production of TV documentaries about her, and a programming language being named after her. This recognition of Ada's special role is largely as a result of her translation of Menabrea's paper—or to be precise, a series of notes she appended to it. It has been said that she worked out programs for the Analytical Engine—this is something of an exaggeration but, in her lengthy notes, twice the size of the original paper, she showed a powerful understanding of the importance of the Analytical Engine and the means by which it would be used.

There seems little doubt that Ada could have helped Babbage greatly, nor was she lacking in ability or enthusiasm, but her approaches to Babbage were turned down in a cold way that reflected the miserably limited idea of women's intellectual capabilities that was common at the time. Babbage was no more sexist than the next man—but unfortunately the next man was sexist indeed. Babbage was very happy to show off his mechanical toys and wonders to "the fair sex," but he did not expect women, and Ada in particular, to make a contribution to the development of technology. She was to have no further input into the development of computing.

From Babbage's engines, real and imagined, it was the punched card itself that now carried forward the torch of automated computing, fittingly to help with that oldest of statistical nightmares, the collation of census results. The man who brought the punched card into practical use in computing was Herman Hollerith.

Born in Buffalo, New York, in 1860, Hollerith was involved in the manual processing of the data from the 1880 census. As overworked clerks plowed through the vast stacks of paper, it became obvious that the whole census operation was in danger of collapse. The amount of information to be handled was growing so quickly that it was feared that, by the next census in 1890, the system would reach breaking point. It seemed entirely possible that it would take more than ten years to process the information collected in the 1890 census. By the time 1900 came around, the whole system would have collapsed in an administrative tangle. Something had to change.

Hollerith, familiar with the Jacquard loom, spotted the potential for using punched cards to hold data. Each card on the loom provided a set of instructions—it was a representation of a series of numbers. Once the data was on a card, it could be repeatedly used in different patterns. A similar card would allow the same flexibility in the handling of census data, if each line of information was punched onto a card. All that was needed was a mechanical device that could sort the cards according to the positions of the punched holes.

Initially this was all that Hollerith's devices did, funneling cards into different bins to be counted by hand. He had devised a way of sorting census data automatically. But with time, the punched card machines would also count the cards and began to tabulate the collected data. By gradual steps, Hollerith's "punched card tabulators" transformed into the mechanical computers of the company that grew out of his original Tabulating Machine Company—International Business Machines, later known only by its initials IBM.

From the development of complex tabulators, the twentieth century saw huge steps forward as mechanical processes were replaced by vacuum tube electronics and then by transistors and integrated circuits. The history of electronic computers is messy. As Mike Hally points out in his *Electronic Brains—Stories from the Dawn of the Computer Age,* it is impossible to say exactly which was the original electronic computer, with a half dozen contenders in the United States and the UK, each achieving some "first." What is clear is that the 1940s and 1950s saw the birth of the true, programmable computer that held both data and program in memory (the von Neumann architecture, named after the great U.S. mathematician and theorist, John von Neumann).

Later, the punched cards themselves were replaced by teletypes, then visual displays (though there are still plenty of people around who have computed with punched cards carrying "Hollerith strings"). But the groundwork had been laid for computers that would become faster and faster, providing vastly improved capabilities. Though the modern computer has no direct lineage that connects it to Babbage's Analytical Engine, it is without doubt the spiritual child of that remarkable concept.

So powerful has the computer become that it is easy to assume that there is no limit to its capabilities, given sufficient time and appropriate hardware and software. David Harel, in his book *Computers Ltd,* quotes from a 1984 article from *Time* magazine that claimed that, with the right kind of software, a computer can do whatever you want it to. There may be limits on a particular machine, the author claimed, but there are no limits to the capabilities of software. Sadly, he was absolutely wrong.

In the first place, it's not just a matter of "there may be limits" on the computer hardware—there *will* come a point where it becomes difficult to go any further. For many years, microprocessors have obeyed the rule of thumb called Moore's Law, stating that they will double in numbers of transistors on the chip on a regular basis. This arose from an observation made in 1965 by Intel cofounder Gordon Moore (then head of R&D at Fairchild Semiconductor), based on just a few years' data, that transistor count, a crude measure of computing power, doubled each year. Moore later revised the frequency with which he expected the computer count to double to every *two* years, and the reality has remained consistently between the two predictions for four decades.

Figure 6.1. Realities of Moore's Law (courtesy of Intel).

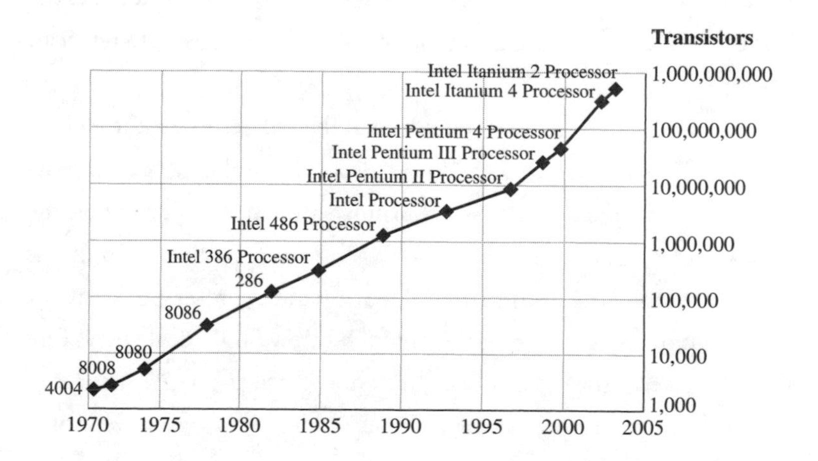

However, there are physical limits that will stop this steady progress from continuing forever. Increasing speed involves increasing miniaturization, and the physics suggests that there are boundaries beyond which this becomes first impractical, then impossible. At some point the computing architecture will have to deal with single electrons at a time, and beyond that there is nowhere smaller to go. This doesn't mean processor power won't keep increasing for many years (at the time of writing in 2005, Intel is predicting fifteen-plus more years before having to make a significant change of tack). Manufacturers have shown themselves to be supremely clever when it comes to finding different ways to cram more into the same space, or to change the topology of the chip so that more space becomes available, for instance by expanding the structure vertically as well as horizontally—but limits will be hit eventually.

Worse still, though, the software that can run on the computer is inherently limited. As long ago as the 1930s, Alan Turing, the mathematical genius at the heart of the wartime Bletchley Park center that broke the secrets of the German Enigma coding machines, proved that there were some problems that can never be resolved by a computer. And there are many more that, though theoretically soluble, would take the best computer longer than the entire lifetime of the universe to reach an answer.

Back in the 1970s, it was thought that the answer to the limitations of conventional computers would be found in parallel processing. Instead of one computer processor churning through the calculations, the idea was that a parallel machine would work on thousands or millions of calculations simultaneously. While such parallel processors are regularly used for specialist problems (though rarely on the scale they were first envisaged), they still didn't give enough extra power to succeed with the really massive problems that would tie up a conventional computer for the rest of time. And parallel processors have also never broken into the mainstream, largely because it has proved difficult to generalize the parallel approach.

The sheer effort of providing a means of splitting general purpose computer applications across the processors soaked up too much of the power. To make really effective use of parallel processors, the code has to be specially written to suit the problem. We are used to general purpose operating systems like Windows, Linux, or Mac OS that take the instructions for our requirements—whether it's browsing the Web or working through a complex mathematical calculation—and hand them to the computer's hardware to process them.

With a parallel computer, the way the task is split between the processors depends on the problem, so it is hard to have a general-purpose operating system. Instead, it proved necessary to tweak the way the system operated depending on the problem, and that is no good for flexible commercial applications. Although there are commercial parallel platforms, like the IBM/Sony/Toshiba Cell chip in the PlayStation 3, they can only speed up simple repetitious actions. Such chips don't provide sufficient change in order of magnitude to cope with the truly immense computing problems.

However, conventional parallel processing isn't the only option to be considered as a solution to the mathematical limitations of traditional computers. For a long time, it has seemed conceivable (if not practical) that computer scientists could make use of the peculiarities of the quantum world to drive the next generation of computer. Paul Benioff, an American researcher, suggested using the spin property of atoms to simulate a conventional computer back in 1979, but we have already met the person who was most responsible for giving the concept of quantum computing a kick start: the remarkable Richard Feynman.

By 1981, when he made his intervention, Feynman was already the grand old man of the physics world. Part of the original atomic bomb team and with a Nobel Prize for his work on quantum electrodynamics under his belt, Feynman had a habit of throwing in rapid contributions in different fields that would have a huge impact, even though he himself might never show any further interest in the subject. It was Feynman who, in 1959, first came up with the idea of tiny, self-replicating machines—nanotechnology—and Feynman again who, in 1986, dramatically demonstrated at the inquiry into *Challenger* disaster that the rubber O rings responsible for the accident lost elasticity when cooled. Being the showman he was, Feynman did this in front of TV cameras, plunging a small O ring into the iced water he had been given to drink.

His contribution to quantum computing was equally quirky. The sixty-three-year-old Feynman had been invited to give a keynote speech to a conference at MIT on physics and computation. His speech, called "Simulating Physics with Computers," which was later developed in a paper for the *Journal of Statistical Physics* as "The Computer as a Physical System," made a startling challenge. Computers had long been used to make approximate models of reality, whether it's a queuing model used to control an automobile production plant or the complex models used by weather forecasters. But Feynman suggested there could be something more—a computer that could absolutely model the physical world (or at least part of it).

The problem with doing this is that at the basic level the world runs on quantum principles, and quantum theory depends on a probabilistic structure. Conventional computers don't understand randomness—they are "deterministic." Theirs is the clockwork certainty of Newton's universe, not quantum theory's statistical predictions. If a computer follows a particular set of instructions, the same thing will happen, time and again. Usually this is a good thing. It wouldn't be helpful if, every time you fired up your word processor, a totally random set of controls, initiating an equally random set of activities, popped up. Nor would it be entirely encouraging if a different value was generated on each occasion that you added the same two numbers in a spreadsheet. But true randomness is an essential if a computer is to accurately reflect the quantum world, and that is beyond the capability of conventional computing.

If you've done any programming, or even used a fairly sophisticated spreadsheet, you might feel by now that I've missed something. Every programming language from Basic on up provides the programmer with a random number module, intended to provide exactly what a quantum simulation is looking for. My Excel spreadsheet, for example, has a function called RAND, which, according to the help system, "returns an evenly distributed random number greater than or equal to 0 and less than 1. A new random number is returned every time the worksheet is calculated." Slip in the appropriate bit of code and out pops a random number. So what's the problem?

The problem is that, useful though this facility is to make a prize draw, or display a series of photos in random order, it's not good enough to model quantum reality. This is because the people who write programming languages have lied to you. That random number function doesn't deal in true random numbers—it produces a pseudorandom number.

Typically, a random-number generator works by taking a rapidly changing "seed" value (say, the number of seconds that have elapsed since midnight on January 1, 1900) and plugging it into a formula, producing results that jump around in a very similar way to a random effect. If you use the same starting point twice in a row, you will get exactly the same result—the generator isn't really random. (This starting-point business is often hidden by the generator's automatically taking this seed from the computer's clock without bothering to tell you.)

It doesn't take a hugely complex calculation to produce an output that twitches around in such a seemingly random way. At its simplest, a pseudo-random number generator might be something like this:

$$\text{Next Value} = (1366 \times \text{Previous Value} + 150899) \text{ modulo } 714025$$

The "modulo" part here just means that the output in this case is whatever remainder is left over when dividing (1366 × Previous Value + 150899) by 714025. This may well be the technique used by the software on your PC, though there are now significantly better algorithms, including the intriguingly named Mersenne Twister, devised by Makoto Matsumoto and Takuji Nishimura in 1996. But whatever technique a conventional computer makes use of, it isn't truly random.

In his 1981 speech, Feynman discussed the possibility of building a computer that would properly simulate quantum effects by basing the workings of the computer itself on quantum mechanics. Just as, in 1930s, Alan Turing had described a universal conventional computer, which would do anything any normal computer could do (though often very slowly), Feynman speculated that there could be a universal quantum computer that would simulate any quantum system perfectly. A conventional computer hasn't the ability to simulate a quantum system properly but—argued Feynman—basing a computer on a quantum system would make it inherently capable of *simulating* a quantum system because it already *is* one.

Whether or not this universal concept could ever be made a reality (the universal Turing machine isn't a practical computer, so there is no reason why Feynman's universal quantum machine had to be), Richard Feynman had lit the fuse of enthusiasm for using quantum effects in the heart of a computer. He, himself, never really took the leap into just how much quantum computers could achieve. Yet Feynman's input helped others to think of enabling a quantum machine to achieve goals that wouldn't be possible with a conventional computer.

There is a potential cause of confusion that needs clearing up here. A "conventional" electronic computer, the ordinary box on your desktop, already makes use of quantum mechanics to work, so what's the big deal? After all, quantum theory entered the mix with electronics. Originally, as we have seen, computers were mechanical, then electromechanical, as the punched card readers used electrical switching and circuitry to manipulate and read punched cards.

Then came vacuum tubes (called valves in Great Britain). These were the first true electronic devices. At the heart of any computer are a series of switches and flow controls. To undertake a calculation using electricity, one electrical circuit has to be able to switch another—that, in effect, is what a vacuum tube does. Inside the tube—a device like a cylindrical glass lightbulb that could be anything from an inch to a foot in height—electrons flow across the vacuum that fills the tube, carrying a current.

In the basic version, a diode, there is a heated filament that gives off the electrons, and a metal plate farther down the tube that they are attracted to. As that plate isn't heated, it isn't spewing out electrons, so the current flow is only one way. In the more sophisticated triode (the vacuum tube equivalent of a transistor), one electrical circuit can control another, more powerful current, switching or amplifying a signal. The important capability of vacuum tubes is that banks of these devices can be used to build the basic logic of a computer.

Vacuum tubes are all very well, but they are temperamental. Their glass shells are delicate. The heater that is used to generate electrons to carry the current can burn out like a lightbulb. And the vacuum inside the tube can lead to its own problems. James Blish, the man behind the story "Beep" in chapter 5, wrote another short story in which probes are sent to explore the atmosphere of Jupiter. Blish comments that these probes can't carry electronics (that's to say vacuum tubes) because the pressure in Jupiter's atmosphere is so high that the fragile tubes, encasing a much lower pressure, would implode.

Blish's timing was bad because, within a year or two of the story being published, the transistor arrived on the scene. And transistors, being solid state, without the need for a delicate glass tube surrounding a dangerously low-pressure interior, would have no problem in Jupiter's atmosphere.

The key to developing transistors and overcoming the fragility of the vacuum tube was discovered in a mixed-up piece of crystal called a semiconductor. In the electrical world, most substances are conductors or insulators. Conductors have plenty of electrons hanging around in the material waiting to move freely from place to place, allowing electrical current to flow (the most obvious examples are metals); insulators, or nonconductors, like plastics, glass, and ceramics, have electrons that are much more tightly bound in place, lacking that freedom to roam.

Glass might be a good insulator but the main element that goes into the making of glass, silicon, is a whole different animal in its raw form. It sits somewhere between a conductor and an insulator. It's a semiconductor. This doesn't meant that it conducts a little bit but, much more interestingly, that it conducts some of the time and it insulates some of the time, depending on a number of factors.

The essential component that stops a semiconductor from being just another insulator like glass is the scatter of impurities it carries. These are tiny fragments of extra elements—in the early semiconductors mostly boron or arsenic—that introduce extra free electrons (or gaps where electrons could go, referred to as "holes"). The simplest form of solid-state device, a solid-state diode, has a portion with intentionally added impurities that increase the free-electron level (the process of adding the impurities is known as "doping"), and a portion with extra holes. If current is applied to the free-electron side, electricity will flow toward the "holey" side; but reverse the current and nothing happens, because the holes fill up and stop conduction.

By the time this type of diode was developed by Russell Ohl, at Bell Labs at the start of the 1940s, semiconductors had already been in use for quite a while, almost by accident. It had been found that by sliding a fine thread of metal (often tungsten, the material most frequently used in lightbulb filaments) across the surface of a crystal such as galena (lead sulfide) or iron pyrite, it was possible to pick up radio signals. These crude radio receivers were called "cat's-whisker sets" because of the appearance of the thin filament of tungsten (they never used actual cat's whiskers). Ohl had started with a series of these cat's-whisker setups and gradually modified the components until he got to the solid-state diode.

By the mid 1940s, Ohl was getting as far as he could alone, and the development was taken over by two other Bell researchers, the practically minded Walter Brattain and theoretician John Bardeen. By Christmas of 1947, the pair were able to demonstrate an early solid state equivalent of a triode vacuum tube.

To even the least commercially minded of those present it was obvious that this was going to be a world class product, and it needed an appropriately catchy name. They toyed with the clumsy "surface state amplifier" and even "iotatron," before one of Brattain and Bardeen's colleagues, John Pierce, decided that the underlying principle of switching currents could be seen in a back-to-front way as transferring resistance (resistance being the ability to hold up a current)—hence the transistor.

What was phenomenal here was not only the capabilities of the individual component but the way it was possible to extend the basic solid state, semiconductor electronics of the diode, first into transistors, where the conductivity of the semiconductor was modified by applying an electrical field, and then into integrated circuits, where whole sets of transistors and diodes could be etched into the surface of a silicon wafer.

The action that's played out inside a semiconductor chip takes place at the quantum level. Arguably, even vacuum tubes are quantum devices because they make use of electrons, but solid-state devices based on semiconductors rely on the peculiarities of quantum effects to be able to function at all. Some semiconductors even change their conductivity when exposed to light, bringing in quantum electrodynamics (see page 72) and making possible such light-sensitive devices as digital cameras.

In reality, then, quantum physics was already part of the mix as soon as solid state electronics came on the scene. Without quantum effects, the transistor and the microchip would not exist. But to go further, to leap to the next generation of computers, needed a bold step of imagination. The realization that the states of fundamental particles—photons or atoms or molecules—could act as the bits at the heart of the computer.

The pursuit of smaller and smaller microchips might inevitably suggest a computer that worked at the level of individual quantum particles, as eventually there is nowhere else to go, but it was innovative British physicist David Deutsch who first spotted that such a computer would be capable not only of being smaller and faster than a conventional computer, but also of undertaking operations that no other computer could achieve. Deutsch had made something of a name for himself in his enthusiastic (in fact, near evangelical) support for the "many worlds" interpretation of quantum theory that had been first suggested by Princeton graduate student Hugh Everett III in 1957.

As we have already seen, the many worlds theory explains the quantum world in all its peculiarity by suggesting that each time any event happens, either the universe clones off new copies of itself reflecting every possible outcome of the event, or there is a complex "multiverse" incorporating all the possible states of the vast probability wave that is the universe; and the reality we experience flips between states every time an event occurs. David Deutsch remains the leading proponent of the "many worlds" interpretation, and managed to come up with an ingenious if highly impractical way of telling whether the multiverse really exists.

Deutsch envisaged constructing a self-aware computer. This is not that old science-fiction favorite, a robot like Asimov's Bicentennial Man with humanlike consciousness, but a simpler machine that is capable of observing its own state. To make it work, Deutsch realized that the machine would have to be a quantum computer, one in which the computational elements were based on quantum objects, because this computer has to observe its own workings and comment on how it "feels" when it observes a particular quantum state, a requirement that would not be possible with conventional computer architectures.

In the more common interpretations of quantum theory, the self-aware computer would only know about one outcome from a measurement but, in the multiverse, all outcomes would be occurring, and Deutsch believes the computer would be aware of every one. This is a subtle concept. Deutsch explains it like this: "What [the computer] tries to observe is an interference phenomenon between different states of his [*sic*] own brain. In other words, he tries to observe the effect of different internal states of his brain in different universes interacting with one another."

Whether this bizarre experiment would in truth prove the existence of the multiverse, it got Deutsch thinking about computers based on quantum elements, and in 1985 he published a paper on true quantum computers. As well as going beyond Richard Feynman's universal simulator of quantum systems, he showed that a quantum computer could do anything a normal computer could do, and, crucially, that it could make use of the peculiarities of the quantum world to provide parallel operations for which there could be no equivalent in a normal computer.

If, instead of the 0 or 1 bits used in a conventional computer, we could use the quantum state of a particle as a bit, we have a part of a computer that can be in more than one state at once: a bit that stores not zero or one, but both simultaneously. At a symposium in London at the end of March 1993, Ben Schumacher, from Kenyon College in rural Ohio, coined the name "qubit" (it sounds like "cue-bit") for one of these quantum state bits.

If there was some way of loading a problem into a series of qubits, and of interpreting the results (as we shall see, neither of these are trivial tasks), we might be able to carry out vast numbers of calculations simultaneously with only a very small processor. Like the massively parallel processor, the chances are that operating system software would have to be individually developed for each special requirement, but the potential return on that effort is so huge with a quantum computer that it would be worthwhile.

It may seem initially that the qubit isn't a very significant step forward. The strange way a quantum particle can be in a superposition of two states gives it two potential values at the same time. Big deal. An ordinary bit can hold one value, so we're only talking doubling its capacity, aren't we? How can this result in a computer that could do things that no conventional computer would *ever* be capable of, as David Deutsch suggested? In the quantum world, things are never as simple as they first appear, and in this instance the complexity is to our advantage.

Take the basic case of the polarization of a photon that was used in variants of the EPR experiment that started the whole business of entanglement. Once we measure the polarization of a photon, that polarization has (in one sense, at least) a specific direction. If we were to measure polarization again in the same direction we would get 100 percent agreement that it was polarized in that direction. If we were to measure polarization at 90 degrees to that angle we would get nothing, again with absolute certainty. Of course, measurements taken in between constitute a different matter entirely, which we will need to come back to, but let's just consider that simple direction in which we get 100 percent agreement.

In effect, the polarization stores, as a piece of information, a direction—an angle in a particular plane. But how would a computer describe that angle in terms of bits? To store the information, we would need to hold a decimal fraction (e.g., 45.33421662 . . . degrees clockwise from the horizontal). Using conventional bits, we would need an infinite string of numbers to describe that direction exactly. Yet *all* that information can be packed into one qubit. You could say that a qubit is analog rather than digital—it can hold a smoothly varying quantity, rather than a quantized value that increases in fixed jumps. More than that, in theory, before measurement, the qubit holds a whole other range of information—the probabilities of having one or other polarization at any angle you choose.

Tim Spiller of Hewlett Packard has a very effective image to get a feel for the difference between a conventional bit with its two-way option of 0 or 1 and a qubit. "Picture it as a classical bit being only black or white, but a qubit having every color you like." That's *every* color, not just the seven in the rainbow, or the millions of colors on a high-resolution computer monitor, but the whole infinite range of possibilities.

It's one thing for our qubit to have all this nominal information stored in it, but we also need to be able to do something with it. That is a thought both tantalizing and frustrating, particularly as you add the information that can be stored in a number of qubits together. It only takes a ridiculously small number of qubits to hold a bizarrely large set of values.

The PC that I'm writing this book on contains 256 megabytes of random access memory—pathetically cramped by modern standards, but enough for most jobs. That's 2,147,483,648 bits in total (a little smaller than you might imagine, as each byte is 8 bits, not 10), but each bit is only capable of being in one of two possible states. By contrast, just 500 qubits can represent more complex numbers, each infinitely detailed, than there are atoms in the universe. If we just held a rounded version of those numbers, taking up 256 conventional bits per number, this would be the equivalent of 32,000 terabytes of information held in those 500 qubits. (One terabyte is approximate a trillion bytes or a thousand gigabytes.)

But there's a big "if" in there—*if* you can make use of that quantum information. Getting information into or out of a quantum computer is anything but trivial. Usually the process will involve tying up several qubits for each qubit actively used, and the use of powerful quantum effects, especially entanglement.

To see why, just imagine a single qubit, sitting happily in a superposition of the 0 and 1 states. If we measure its value, it will always come out as 0 or 1. The quantum state will provide the probability of getting a particular result, but a measurement can only ever result in an outcome of 0 or 1. If we stretch Tim Spiller's black-and-white-versus-color metaphor, it's a bit like watching the world on an old black-and-white TV. We know that world is really in spectacular color, but we can only ever see it on our screen in black and white. To read out the results of a quantum computer, we have to peer into that full-color operation in frustrating monochrome.

This might suggest that the whole business of quantum computing is a mirage—it sounds good but, once you try to use it, once you peer into the black-and-white TV, there's nothing there to make use of. Luckily it's no mirage. It *is* possible to engage the power of quantum computing and still get results out—it just isn't simple. The most trivial way to get around the situation is to have a very complex problem whose answer is much simpler—perhaps yes or no. We never need to see the complexity of the actual calculation (in fact, we can't see the detail without destroying the whole process), but we can see the result. In other cases, entanglement may be used to link together different parts of a quantum computer and produce a more complex output—but we should never overlook the difficulty involved in this challenge.

In many ways, there is a parallel between the state of quantum computers in the early years of the twenty-first century and the mechanical computing engines that Ada King imagined using over a hundred years before. We know that we can do a lot with a quantum computer—more than is possible with any conventional machine. We even know *how* to program some otherwise impossible calculations, if only we had an actual quantum computer to work with. But, as yet, we can't build the things. So dramatic is this dichotomy between theory and practice that quantum computers have, without doubt, had more written about them by serious scientists than has any other nonexistent technology.

The simplest achievement we know is possible in practical terms is to pack two whole bits of information usably into a single qubit—this is given the overdramatic title "superdense coding" by quantum computer scientists. As long as we could get a handle on those infinitely long numbers in the heart of the qubit, this possibility shouldn't be a surprise. In fact the result was effectively proved over a hundred years ago by a master of the infinite, the German mathematician Georg Cantor.

Cantor was a genius who was failed by the system. He should have been a meteoric star in the German academic firmament in the last years of the nineteenth century—instead he was held back at the University of Halle. If Cantor had been a musician, this wouldn't have been a bad thing, as Halle had a great reputation for music. Sadly, though, it was second rate when it came to math. To make matters worse, Cantor was prevented from publishing in the most respected journals of the time. While it's true that Cantor had some strange ideas, these alone would not have suppressed his work and his career for long. Instead, his imaginative work was widely criticized as a result of a long feud with well-established mathematician Leopold Kronecker.

Kronecker was nowhere near as original a thinker as Cantor, but he had superb political connections. Initially Cantor's mentor, he took against his protégé, first baffled and then horrified by Cantor's ideas on infinity. Kronecker was a man who only really believed in whole numbers and fractions that could be made from the ratios of whole numbers. He didn't accept that "real" numbers—decimal fractions that go on forever and which need not have a representation in terms of a ratio of whole numbers—existed. Yet Cantor's whole scheme of the infinite was based on real numbers. Cantor demonstrated, for example, with a blisteringly simple proof, that there were more fractions between 0 and 1 than composed the infinite set of whole numbers. According to this discovery of Cantor's, it was possible to have one infinity bigger than another—a concept that positively disgusted Kronecker.

This is such a remarkable result—and the proof is so stunningly simple—that it is worth digressing for a moment to see how Cantor achieved it. What is described below is a slightly simplified version of the original, but it puts across the sheer power and simplicity of Cantor's thinking. Imagine we decided to produce a table in which we wrote down every number between 0 and 1—every single decimal fraction that exists. This is obviously not possible in practice as the table would be infinitely long, but imagine doing so in principle.

If it were possible *even in principle* to produce such a table, Cantor could prove that the size of this set of numbers was the same as the size of the set of whole numbers, the integers. One of his first major mathematical breakthroughs had been to show that two sets of numbers are the same size (or, they have the same cardinality, in math-speak) if you can work through one of the sets and pair off each member with an equivalent member of the other set. Imagine, for instance, pairing off the legs on a dog with the horsemen of the Apocalypse. Even if you didn't know how many there were of either, you would find out the sets were same size by pairing off one leg with each horseman, and having none left over.

Cantor realized, that if it *were* possible to have a table containing every fraction between 0 and 1, then he could work through the entries in this table one at a time and pair each off with one of the integers. That would prove that they were the same size. It seems an obvious result. Yet Cantor found a way to demonstrate that such a pairing off could never be achieved.

Imagine that we scramble up his table of fractions between 0 and 1 into a random order. This is simply so we can represent the first few entries in the table on paper. Otherwise the first entry would be 0.000 . . . all the way to infinity, the second entry would be 0.000 . . . all the way to infinity with 1 as the final digit, and so on—to say the least, impractical to represent in a finite book. Here's the start of a scrambled table:

0.4850252988....
0.0953822319....
0.7773943015....
0.3864932203....
0.5298932641....

Figure 6.2. Cantor's randomly ordered table of numbers between 0 and 1.

Once we've got these numbers, we can see Cantor's genius at providing a simple solution to what seems a very complex problem. He made up a new number by taking the first decimal place of the first number, the second decimal place of the second number, and so on, through the table (the digits shown in bold in the table above).

In this case, the number is 0.49749 . . .

He then added 1 to each of the digits (adding 1 to 9 resulted in 0) to produce a new number, 0.50850 . . .

Now this is a remarkable number. It clearly isn't the first number in the table, because the first decimal place is different. That's how we defined the new number. It isn't the second number in the table, because the second decimal place is different. And so on, throughout the table. We have just generated a number that isn't in the table. Anywhere. Cantor had shown with this wonderfully simple proof that it wasn't possible, even in principle, to produce a table that contained every fraction between 0 to 1. There were too many numbers between 0 and 1 to be able to put them in one-to-one correspondence with the integers, more numbers than the whole infinity of integers. Cantor demonstrated that the continuum of numbers from 0 to 1 represented a new infinity, bigger than the basic infinity.

Once Cantor had produced this stunning result, using a proof that effectively requires no mathematics, he went on to extend his workings beyond a single dimension. The numbers between 0 and 1 can be thought of as all the points in a one-dimensional line, a number line—imagine a ruler, marked off with every number between 0 and 1. Cantor wanted to extend his working to the number of points in a plane, or a cube—or a structure of any number of dimensions you'd care to describe. We all learn at school that to identify a point on a flat surface we need two numbers—the coordinates. These are usually called Cartesian coordinates after the French philosopher René Descartes, though the basic concept that you could represent a point on a map by two measurements was known long before Descartes' time. What Cantor showed with remarkable simplicity was that we've all been told a lie. You don't need two numbers to represent a point on a plane (a map reference, for example); you only need one.

Let's say we have a point that is represented as 0.53419 on the X axis and 0.82948 on the Y axis. (In math, the horizontal axis is traditionally labeled X, and the vertical axis labeled Y, for no good reason.) Cantor showed that such a point could be represented uniquely by a single number. Let's make that Y axis coordinate bold: **0.82948.** Now all we need to do is alternate digits between the two numbers to produce a single number, 0.**8**5**3**2**4**1**9**4**1**9**8**, which uniquely represents that point. Of course, you could say we're cheating because the new number is twice as long as the old one. But it is still only a single real number—and we only needed that one number to represent the position.

Even if the X and Y coordinates were infinitely long, we could still produce a single infinite number that precisely specified the values. Due to the impracticalities of dealing directly with infinity, that

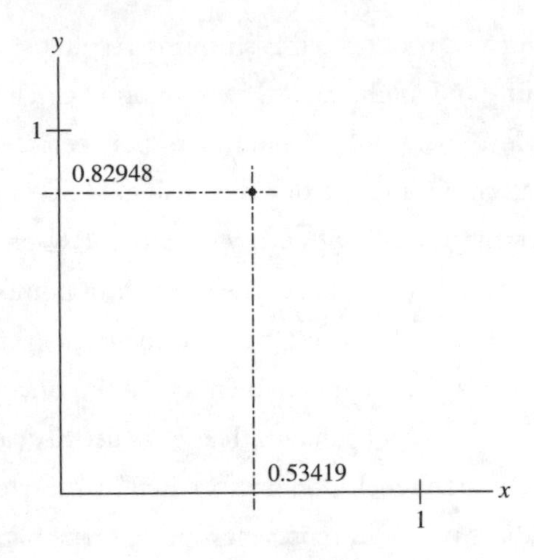

Figure 6.3. Representing a point in two dimensions with Cartesian coordinates.

could get rather complex, but here's a simplified version: if the X coordinate is 0.333333 . . . (where . . . means it goes on all the way to infinity) and the Y coordinate is 0.777777 . . . , then the single number is 0.373737. . . .

If there were some way of getting to, and modifying, the infinitely long numbers that lie at the heart of the qubit, it would be trivial for a qubit to carry two bits of information, or in fact as many bits as you would like. (Cantor's argument works just as well for three dimensions, or for that matter *n* dimensions—just interlace the different coordinates.) Superdense coding doesn't achieve that, but it does enable us to use more of the qubit's capacity than a simple 0 or 1 value—in fact enough to store the information that normally takes two bits to store, giving an effective choice of 0, 1, 2, or 3.

The process involved is a little messy but not that difficult to follow in principle—and it uses entanglement. We start off with two entangled photons. The part of the computer that wants to send two bits of information in a single qubit takes one of the photons, which is to act as the qubit, and processes it. For each different value required, the qubit is put through a slightly different operation. To get 0, nothing is done to it. To get 1, it is put through one sort of quantum gate; to get 2, through another; and so on. We'll come back to what a quantum gate is in a moment—for now, just think of it as a special black box that modifies a qubit in some way.

Given a single qubit, you can't read off which gate it was put through, and so make use of its four possible values. But combine measurements on the qubit with an input from the entangled photon and it's possible to distinguish which state the qubit was put into. Although two photons are used in the measurement, only one was processed in the first place and forms the qubit that needs to be sent around the computer's system—the other photon, the entangled carrier, is inactive until required, and could have been sent to its final destination long before the qubit is ever put through the appropriate gate and dispatched around the computer.

In fact, even the "buy one, get one free" capability of superdense coding has proved harder to achieve in practice than it is in theory. When a group at the University of Innsbruck, including Anton Zeilinger, the most persistent of experimenters in the world of quantum entanglement, put together a system to demonstrate the superdense qubit in action, they found that two of the four states were indistinguishable—instead of getting four values into a qubit, they only got three, what has sometimes been called a "trit" to distinguish it from a traditional, two-value bit. This doesn't mean it isn't possible to achieve the full four values, but the result has so far eluded experimenters.

If superdense coding were the sum achievement of quantum computing it would be an interesting idea, but hardly earth-shattering. However, there are also powerful algorithms that could be programmed into a quantum computer today (if only we had one to program). Algorithms that could totally transform the way we deal with information.

An algorithm is a mathematical tool that can be much simpler than the name sounds: it's just a recipe for producing a result in math. An algorithm consists of a set of rules that operate on some kind of input of data to produce an output. An algorithm can be as basic as the mechanism that we all learn in elementary school for multiplying two numbers together, or something as sophisticated as the algorithms that airlines use to decide how to pack different items in a freight container to maximize the use of space.

In 1996, Lov Grover, of Lucent Technologies' Bell Labs, came up with a method that would drastically speed up a search through a database of information that hadn't been sorted (usually a nightmarishly slow operation). His method was unexcitingly described as "A Fast Quantum Mechanical Algorithm for Database Search" when first mentioned at the ACM Symposium in 1996, but Grover would later show an unusual sense of humor in the title of his paper on the subject for the generally austere *Physics Review Letters*—"Quantum Mechanics Helps in Searching for a Needle in a Haystack."

This quiet sense of humor is typical of the very quiet-spoken man himself. Lov Grover was born in the small Indian town of Roorkee and studied in Delhi to be an electrical engineer. This was still his intention when he emigrated to the United States but, by the time he had taken his PhD at Stanford, Grover was as interested in the purer problems of physics as the practicalities of engineering.

In 1994 Grover joined Bell Labs, one of the few commercial organizations where he could have the freedom he would need to spend time on a quantum algorithm. As Grover admitted when interviewed, "I was lucky to be working in a company like Bell Labs. Even during the crisis that Lucent [Bell's parent company] went through, we were still doing forward looking research . . . and forward looking research is still appreciated. Perhaps not so much as twenty years ago, but still appreciated."

Grover heard about the idea of quantum algorithms at Bell Labs, and, in his own words, "almost immediately got the search algorithm." What Grover had realized is that a quantum computer was uniquely well set up to search for the proverbial needle in a haystack. Or perhaps something slightly more realistic—haystacks lack the structure of even an unsorted database. Let's imagine instead that we have a huge chest of drawers. One, and only one, of the million drawers in our chest contains a needle. How are we going to find it? With a purely random allocation of the drawer, there is no better technique to find the needle than opening the drawers one after the other. And in the worst possible case, if things went as badly as the legendary Murphy's Law suggests, we might have to open 999,999 drawers before getting to the one with the needle in it.

Exactly the same thing applies when searching through an unsorted database—each record of the database corresponds to one drawer. The record holding the information we want could be anywhere in that database. Generally a carefully set up computerized database contains indexes that *are* sorted—but there are plenty of real-world examples of unsorted databases around. Take a random phone number and try to find whose number that is in a conventional paper phone directory, if you want to experience the joys of working with unsorted databases. It's no fun. And despite all the wonders of electronic searching, most of the data in the world isn't in nice, ordered databases, but in unstructured muddles.

This is a seemingly insoluble problem. It doesn't matter how clever your programming of a conventional computer is, it will still have to work through each entry in the database one at a time, hunting for that virtual needle. Okay, your search program could get lucky and find the right record the first time, but the chances are a million to one against. On average, the computer is going to have to check 500,000 of those entries to get to the answer—in the worst case, 999,999. But the quantum world, as we are coming to expect, is very different. With some nifty math, Grover showed that a quantum computer could get away with only searching through a set of items equivalent to the square root of the size of the database. For a database with a million items, that's just 1,000 checks—potentially 998,999 less than working through the whole thing.

This isn't just useful for searching conventional databases, though there's plenty of scope there. As more and more rich but unstructured information becomes available to us, it may well take a quantum computer to power the descendents of search engines like Google, sifting through not just the World Wide Web but every document you've ever accessed, and every library in the world. Just how relevant Grover's work is (if a quantum computer is ever built) was emphasized when, in 2000, he came up with a second algorithm that showed how a quantum computer could achieve the same sort of remarkable speed-up for the sort of vague search that is common in the real world, especially when relying on human memory as a starting point.

Grover gives the example of finding someone in a phone book. Maybe it's someone you met the other day. You can remember that his first name is John, and that he has a common surname, but cannot recall exactly what that surname was. Say, you think he's a Smith with a

50 percent probability, Jones with 30 percent, and Miller with 20 percent. You also remember that he said he could see Broadway from his apartment, and that you noticed the last digits of his phone number are the same as your doctor's number. With just these sorts of fuzzy information, a typical starting point when attacking unstructured real-world requirements, Grover's new algorithm enables a quantum computer to home in on the right result vastly quicker than anything possible with a conventional search.

There are plenty of related problems that are impossible for a conventional computer to crack in any sensible time, but that variants of the same quantum algorithms would make easy. Devising the timetable for a large institution is currently impossible to perfect, as is finding the shortest route to connect a series of destinations on a complex road network. This might be a shock to a user of a product like Microsoft's Streets & Trips that claims to find the best route for any journey.

In order to come to a solution of this sort of problem, programs like Streets & Trips have to rely on estimation. High-class guesswork, if you like. This isn't as bad as it sounds. It's like estimating the length of a piece of wood by how long your thumb is. You can get to a reasonable approximation by measuring out thumb lengths—in fact your guess will almost definitely be right to the nearest thumb (if the plank isn't too long). Similarly, and more precisely, if the methods used by route-planning software to come up with trip directions produce an answer at all (and, statistically, there will be some routes that they'll never find appropriate directions for), they can be shown to get within a certain range of the best answer. But such solutions remain approximate. Developments of Grover's algorithm would make some of these fiendishly complex problems completely soluble.

Other quantum methods have been devised to manage equally time-consuming but important mathematical feats; most important of these being producing the factors of a large number. In fact, these two applications—fast search, and factoring large numbers—are between them the big hopes for quantum computing, though there may well be many other uses for quantum computing that have yet to be devised. According to Lov Grover, "Not everyone agrees with this, but I believe there are many more quantum algorithms waiting to be discovered."

Although quantum searching has the most obvious value for today's users, the realization by Peter Shor, of AT&T, that quantum computers could vastly speed up the breaking down of huge numbers into their factors was the discovery that brought quantum computing to the top of many computer scientists' agendas, not so much with enthusiasm as with fear.

Peter Shor was born in 1959 in New York City. After gaining a degree at Caltech and a doctorate at MIT, he took a job at AT&T Bell Laboratories. For several years, he worked on algorithms for conventional computers and on probability. But when, in 1994, Shor came up with an algorithm that would enable a quantum computer to crack factorization in a way that no conventional computer could match, it sent shock waves through the computer science community. This was something they had always hoped that it would *not* be possible to achieve.

Let's go back to that picture at the start of the chapter of our brave new, Internet-linked, computer-driven world. In such a world, we all have a burning desire for more open information. Wouldn't it be great if every book ever written, every academic paper, every newspaper, and every magazine were all instantly accessible online? Some of that is happening now. The company behind the search engine Google, for instance, intends to digitize millions of books with the help of university libraries, while more and more journals and documents are available online in a few clicks of the mouse. It would be a devastatingly superb resource (and probably would require some of that superspeed searching from Grover's Search Algorithm to be used most effectively) if everything ever published were available online. Such open access would give ordinary people the huge power that the ability to read, compare, and collate crucial information gives. Yet this picture of true freedom of information has its dark side.

Would you want every piece of information everywhere accessible to everyone? The knee-jerk reaction might be, "Yes, I've nothing to hide, so why should anyone else?" but the reality is a resounding no. As we saw in chapter 4, we all have secrets, such as our credit card numbers, that we want to keep hidden. Others want to protect the copyright of their work. And the one thing that stands between an Internet pirate and that restricted information is the difficulty of deducing the factors of very large numbers, one of the exact fields in which quantum computers can run rings around any conventional machine.

Whenever you send an encrypted e-mail, or see a little padlock at the bottom of your Web browser indicating a secure connection as you tap in your credit card number, you are making use of one of the cleverest bits of secrecy technology ever devised. As we saw in chapter 4, the most secure way to send a secret message is to use a one-time pad private key. If the key is as long as the message itself, and the information in the key is truly random, the message is impossible to break. Totally impossible.

The problem then becomes distributing that key without the key itself being intercepted—in effect the key becomes a secret message in its own right. This problem can be overcome if there were such a thing as a public-key/private-key system. In such a system, the key used to encode a message is known to everyone—so it can be published everywhere and anyone can use it to hide the contents of a message.

But the key used to decode that encrypted information, a different one from the public key, is only to be known to the recipient. That remains private.

Imagine I had a special, self-sealing material that could only be cut with a weird knife. I'm the only one with that knife. I can give piles of the material to everyone in the world. If they want to send me a secret message, they just wrap it in the special material. Anyone can do that. But because I own the only weird knife in existence, I'm the only one who can open the package and read the message. The public key is like the special material; the private key is my weird knife. Of course I have to give out the special material first, but that's no secret—anyone can have it—and once I've given it out, I can receive information securely. To send it back, we have to have the opposite setup—I use someone else's freely available special material, and the other person has his own, different weird knife.

As a solution, this seems as much use in security terms as waving a magic wand and chanting a security spell but, in 1977, three men working at MIT's Laboratory for Computer Science realized that there was a practical way to make the magic come true. The technique they developed, known as the RSA algorithm after their initials (Ronald Rivest, Adi Shamir, and Leonard Adleman), and variants on that algorithm, are at the heart of all modern high-level computer-based security.

As an aside, this should probably be called the CRSA algorithm, as the method was originally devised three years earlier in GCHQ, the Cheltenham-based British intelligence center, by Clifford Cocks. Unfortunately for Cocks, he was working for his government, and the algorithm was considered important to national security, so his discovery was kept secret and was not publicized until 1997, when it was much too late to do anything about the name RSA (or for that matter, about patents and royalties).

The process of the RSA algorithm involves a number of stages of large-scale computation, but it contains nothing beyond the capabilities of the typical desktop computer. The would-be recipient of a secret message takes two huge prime numbers (numbers that can only be divided by themselves and one) and multiplies them together. The resultant, even huger number, along with a second, randomly chosen number is shared with the world, but only the recipient knows the original primes. The secret information is encoded, using these values in a way that is practically irreversible without knowing the two primes that first went into the equation. So anyone can do the encrypting, but only the receiver can untangle the message and read it.

There is, of course, a flaw in this system. It certainly isn't unbreakable like the one-time pad's key. To be able to decrypt the message, all you need to know apart from the public key is which two primes multiply together to produce the big part of that public key. In principle it is always possible to break the RSA code. If the final number used had been small, cracking the code would be trivial. Say the key was 15; it doesn't take a supercomputer to work out that the two primes in question are 3 and 5.

But RSA makes use of numbers that take up anything between 64 and 4,096 bits of binary space. It could go on forever, but at the moment 4,096 bits is considered pretty safe—bear in mind that you can fit a number bigger than 1 with 1,200 zeros after it into 4,096 bits. That's as many 0s as letters on a page of this book. It doesn't matter how fast computers are at a particular time in their development; it's possible to make the key vast enough to take thousands of years of computing to break it down into its prime factors. But that assumes that we are dealing with a conventional computer.

Enter the quantum machine. Bearing in mind that finding factors that ordinary computers can only dream of is one of the proven capabilities of a quantum computer (should one ever be built), the very idea of a quantum computer brings on nightmares in the computer security field. Of course there are many positive applications of such high-speed factoring, and it's also true that quantum encryption could supersede RSA-style encryption, but the ability to run quantum algorithms and break through public-key encryption would have a devastating impact on the online world.

Perhaps the biggest problem for those facing up to the impact of quantum technology on computer security is a lack of symmetry between the criminal and the person to be protected. All the signs are that quantum computing technology will be expensive and complex, at least to begin with. So it's not going to be available to everyone. But the ability to break through encryption need only be available to a few to cause havoc. In order to protect against this, every user—and we're potentially talking billions—would have to have quantum encryption to be safe. While quantum encryption is a lot cheaper, more practical, and everyday than the idea of a quantum computer, it is still nontrivial to apply.

It's not all bad news, though. Artur Ekert, who devised entanglement-based encryption, believes that quantum computers have much more to offer than has so far been imagined. He takes a very low-level approach, returning to the early concepts of quantum computers as devices that can go way beyond the capabilities of traditional computing. He points out that computers are physical objects and, as such, may be able to solve problems that are mathematically impossible to handle.

Ekert uses a simple example. Imagine two separate rooms, each totally sealed, with a single door and no windows. In the first room are three lightbulbs. In the second are three switches.

One switch corresponds to each bulb. The problem is, you are allowed only one visit to each room (after which the door is shut) to discover which bulb corresponds to which switch. Ekert suggests that providing a purely mathematical solution is infeasible. But a physical approach makes the problem soluble. Enter the switch room, leave switch 1 on for five minutes, then switch it off and turn on switch 2. Immediately visit the bulb room. The bulb corresponding to switch 2 will be on. Of the other two bulbs, the one connected to switch 1 will be warm, while the bulb for switch 3 will be cold. Ekert's challenge is to find other ways that quantum theory and entanglement can enable physical solutions not available to pure math.

Whether or not Ekert's extension into "real-world" computing is possible, should a quantum computer appear by magic on the desk of a computer scientist, its ability to factor large primes would have both a real potential for causing chaos and the capacity to transform the business of searching.

"Appear by magic" is about as far as we've got right now. There is no such thing as a quantum computer to run these clever programs on. Just as Babbage was defeated by the limitations of conventional mechanics, those who want to build quantum computers face a stupendous battle against the complexity of quantum mechanics. At the time of writing, there still is no quantum computer, not even in the controlled environment of the laboratory. But, equally, as Babbage was able to demonstrate a small part of his Difference Engine, so small parts of quantum computers have been constructed, and there is now much more hope of getting to a real working model than there was only a few years ago—thanks to entanglement.

We have already seen that entanglement is needed to make superdense coding possible. In fact, there seems to be no way to have a working quantum computer at all *without* making use of entanglement. To see why, we've got to look at how normal computers work on a theoretical level, and discover why the quantum mechanical nature of qubits insists on getting in the way of the essential working mechanism of computation.

Stripped of all the complexity, working away under the fancy graphical user interface, operating system, and applications, most of what a computer does comes down to taking a binary value—0 or 1—and either copying it or putting it through a "gate." This is a device that *may* modify the value based on the value of the bit itself, or control information that usually comes from another bit. A simple physical equivalent of a gate is a child's shape sorter. If we have a selection of different shapes representing different inputs and try to put them through the square hole, something comes out the other side (1) if the input is square, but nothing comes out (0) if the input is anything else.

It really doesn't matter how the gates work in a computer; it could be anything from an actual physical mechanical gate to an electronic component like a transistor. Probably the most basic useful gate is a NOT gate—this simply says, "Wherever you see a 0, output 1; wherever you see a 1, output 0." A physical analog of a NOT gate would be a mirror, using a left-handed glove as a 0 and a right-handed glove as a 1. Whichever glove you put in front of the mirror, its reflection would fit the other hand, as if it had been passed through a NOT gate.

The NOT gate acts on a single bit, but many of the more powerful gates require at least one more bit as input. For example, the AND gate combines the values of two bits. If both bits are 1, the result is 1; otherwise, for each of the three possibilities of 0 and 0, 0 and 1, or 1 and 0, the result is 0.

Quantum computing has to deal with the same kinds of requirements as a conventional computer but, because the equipment has to deal with fragile qubits, the operations that quantum gates apply tend to be more complex. Even that simplest of actions, copying a value, is not trivial. As we will see later, thanks to the biological sounding "no cloning theorem," a quantum computer can't make a copy of a qubit. However, there is a way around this. Entanglement does make it possible to copy a quantum object as long as you destroy the state of the original object in a method called "teleportation," which we will cover in the next chapter.

Over and above the irritating complexity introduced by the inability to clone (because a lot of conventional computer operations assume that you can take a copy of a value), quantum gates have to be more complex than traditional computer gates because they are operating on multiple states simultaneously. All a NOT gate has to do is switch 0 to 1 and 1 to 0. The quantum equivalent of a NOT gate, an X gate, has to take a qubit that has a certain probability of having the value 0 and a different probability of having the value 1 and swap those two probabilities, a more tricky requirement to manage.

Complex though they sound, such gates are theoretically possible to construct, and have already been used to process qubits, as when a qubit was put through one of four processes in the superdense coding experiment (see page 177). One process is "do nothing," another is an X gate, and so on. Quantum gates have been designed to perform pretty well everything that a quantum computer should require, and in many cases have been built and tested on a very small number of qubits.

The state a qubit is in—before or after passing through a gate—is often represented by a special symbol. It isn't necessary to use these symbols in order to understand entanglement or quantum computers, so they don't generally appear in this book, but should you read more widely on quantum computers (or other aspects of quantum theory) you will inevitably come across it, so it is worth briefly describing them. Say we've got a qubit that has the possibility of being in one of two states, 0 or 1. The "state function"—the mathematical description of the qubit's superposed state, with its different probabilities of having either value if measured—is usually shown like this: |0> to represent the 0 state and |1> to represent the 1 state.

The advantage of using this notation is that it makes clear that we are dealing with something that describes the quantum object's state, not an actual, fixed value of 0 or 1. When a qubit is in a state that is a combination of the two states |0> and |1>, it is written as |0>+|1> for the condition where there is a 50:50 probability of either state appearing when the object is measured, or with extra factors such as a|0>+b|1> to allow for the probabilities of the two outcomes being different.

This notation, which looks very odd to the ordinary eye, is a standard in the quantum world—and it makes more sense when seen in the context in which it was originally envisaged. It was introduced by the British physicist Paul Dirac, who rationalized a lot of the work on quantum theory and the |anything> item is technically called a "ket." This was a feeble joke on Dirac's part, as there is an anti version of it called a "bra" that looks like this: <anything|. Together they make "bra" "ket"—a bracket. These terms tend to be used less these days (the term "bra" for an item of underwear emerged in 1936, shortly after Dirac first used the terms, and the resultant tendency to produce sniggers from an undergraduate audience might have been responsible for the less frequent use of the terms), but the symbols themselves are an important standard for those working in the field.

Quantum computers can seem a little like a wonderful optical illusion, so convincing that we keep reaching for the apparently solid object, only to have it dissolve in front of our eyes. At the time of writing in 2006, although there has been a lot of progress made, we are still a decade or two away from being able to build a practical quantum computer. Lov Grover, who devised the Grover Search Algorithm, is optimistic, though. "It looks a lot more feasible now that in it did ten years ago. Then most people thought it was a theoretical result with no practical implementation. This is no longer the case."

That's encouraging—but how likely is it that that quantum computing will be achieved? Bearing in mind both the difficulty of assembling the quantum computing equivalent of electronics, and the complexity of programming a quantum computer, is it any more than another overhyped, seemingly never-to-be-delivered piece of technology like personal rocket belts or intelligent humanoid robots?

It's arguable that there wouldn't be such an academic industry built around quantum computing if it were just another potentially nice but practically unachievable technological idea. More important, we have to remember that one of the reasons that quantum computing is more important than this (and also why it can be so hard to understand) is that it *isn't* a technology at all; it's something more fundamental. As Michael Nielsen and Isaac Chuang put it in their *Quantum Computation and Quantum Information:*

> *It is tempting to dismiss quantum computation as yet another technological fad in the evolution of the computer that will pass in time, much as other fads have passed—for example the "bubble memories" widely touted as the next big thing in memory during the early 1980s. This is a mistake, since quantum computation is an* abstract paradigm *for information processing that may have many* different *implementations in technology.*

The reason so much work has been put into quantum computers without them even existing is that, provided quantum theory is correct—and there is yet to be the slightest indication that it fails to correspond to experiment—most developments in quantum computing are entirely separate from the technology. Quantum computing and its benefits can be described with math and a knowledge of quantum mechanics alone. The technology to make it a physical reality may take a long time to perfect, but it will come. This is likely to take ten to twenty years, or even a more pessimistic one hundred years, but the clock is ticking. And when usable technology does arrive, quantum computers will be the best-prepared invention ever to hit the production line.

The challenges facing the builders of a quantum computer are not insignificant. These quantum engineers need to construct a series of qubits and link them together. They will have to get information into the system in the first place, then trigger the operation of the computer, and, finally, get the result out. None of these stages are trivial when working at the quantum level. It's as if you were trying to do a complex jigsaw puzzle in the dark with your hands tied behind your back. Pretty difficult—but if the exercise has value, and quantum computers certainly do (unlike doing complex jigsaw puzzles in the dark with your hands tied behind your back), someone, somewhere is liable to come up with a way to do it.

One of the problems that scientists working in the field come up against time after time is that, although there have been a half dozen different mechanisms tried out to provide qubits, each with useful features, each also has significant drawbacks. Quantum computer scientists are a bit like the first of the three little pigs who tried to build his house with straw. The design may be great, but there are problems with the fundamental units of construction.

Not long ago, the first obvious problem would be managing to deal with a single quantum particle at a time. Quantum objects are, after all, incredibly small and difficult to pin down. It's not enough to just generate photons, say. The simple act of turning on a 100-watt lightbulb will produce around 100,000,000,000 photons every billionth of a second—but that profligacy of generation is a problem rather than an asset. Imagine we had a conventional computer with billions of traditional bits. Great. But now imagine that we can't identify a specific bit. Disaster—how do we ever get information in or out? Similarly, a great cloud of photons isn't much use in a quantum computer. We need to be able to generate, manipulate, and detect a single photon at a time.

Before lasers, this was almost impossible, which is why there were real problems with the early attempts to demonstrate entanglement and Bell's theorem in action. As we saw in Aspect's experiment on Bell's theorem, it was the laser that made it possible to work with individual photons. Lasers, unlike ordinary light sources, can have their output attenuated to practically nothing, generating mostly single photons. And the down conversion method now widely used in entanglement makes it is possible to pin down individual photons by detecting (and destroying) their entangled twins. When a twin photon is found, a gate opens to let out the single remaining twin, producing a controlled stream of single photons.

Quantum computers don't have to be based on photons, though. Atoms also could form the basis for qubits. On the atomic front, it was in 1980 that Hans Dehmelt, of the University of Washington, succeeded in isolating a single barium ion (an ion is an atom with electrons missing, or extra electrons added, giving it an electrical charge).

The ion was held in place by electromagnetic fields. The ion's positive charge responds to the field rather in the same way that a magnet can be made to float over other magnets (though the ion has to be boxed in by several fields to prevent it from flying away—the magnet is held back by the opposing force of gravity). Incredibly, when illuminated by the right color of laser light, the single barium ion was visible to the naked eye as a pinprick of brilliance floating in space.

But let's stay with photons for a moment. After all, using light photons as the qubit is probably the simplest of the options available for would-be quantum computer engineers. Here the polarization of the photon, or the phase difference of a number of photons at a particular point (phase being the position on the "up and down" of the light when it is seen as a wave), can provide the information storage.

Photons are plentiful—this isn't a resource that is going to run out easily. And photons are stable, as witness the fact that we can see, with our bare eyes, photons that have traveled from distant stars. Before these photons are destroyed in the interaction with atoms in your eye, they may have been on the journey across space for millions of years. What's more, many of the gates required to operate on photons amount to little more than different combinations of beam splitters (remember, at its most basic, a beam splitter is just a piece of glass or part-silvered mirror). And the results of a photon-based computation can easily be read using optical detectors.

But the ease of obtaining and performing some operations on photons isn't the whole story. Photons don't like to stay in one place and they don't interact with each other very easily. For everyday life, the lack of interaction is just as well. Look through the air in front of you at the moment. Although they're invisible, the air is absolutely teeming with photons flying in all directions: The light that illuminates everything around you and makes it possible for you to see. Hundreds, if not thousands, of radio, TV, and cell phone signals. Wireless networks and garage door remotes. High-energy electromagnetic rays from the depths of space. If photons did easily interact with each other, there would be countless collisions and everything depending on those photons, from the TV broadcast to your ability to see, would stop working in a confused mess of interaction, like a three-dimensional pool table with billions of balls shot into it from all directions at once.

The other photon-wrangling problem, keeping them in one place to be able to make use of them in a computer, is probably less of a difficulty than it first seems. The reason why it's not easy to pin down a photon in one place is obvious when you consider how Einstein came up with special relativity (see page 130). He realized that you couldn't travel alongside a photon, making it stop as a result of relative motion, because the delicate dance of electricity supporting magnetism that supports electricity (and so on) would only work at the speed of light. In the normal state of affairs, photons simply can't stop moving.

Many of the early ideas of trapping a photon involved keeping it in a tiny reflecting box, so it still moved at light speed, but bounced around in a very small confined space. Recently, though, something much more effective has proved possible. You have to bear in mind that the speed of light isn't really a constant. The familiar value of around 186,000 miles a second is its speed in the vacuum of space. That *is* fixed, but as soon as light travels through a material, the interaction between photons and the atoms of matter around them slow the light down. In air, it's a bit slower than in space. In glass, it's dragged all the way back to around 124,000 miles per second. And that's just the beginning.

In 1998, researchers managed to slow photons down to walking pace in a strange form of matter called a Bose-Einstein condensate (physicists have to be having a particularly good day to come up with a snappy name like "photon" or "quark"). This is the fifth state of matter. We are used to matter coming in three forms—solid, liquid, and gas. Since the 1920s, a fourth type of matter has been known, generated when atoms are exposed to sources of great energy like the raging nuclear furnace of the sun—plasma. This is the stage beyond a gas, where the easily removed electrons have been broken off the atoms and the result is a soup of ions—atoms with some electrons missing—and the electrons themselves.

The four states of matter—solid, liquid, gas, and plasma—have a startling parallel in a theory developed over two thousand years ago.

The Greek philosopher Empedocles thought that everything was made up of four elements—earth, water, air, and fire—each bearing a neat similarity to one of the modern states. Some of the ancients thought this wasn't enough. The four elements were limited to the impure "sub-lunar" region, within the orbit of the moon. Outside, it was thought, there should be something more perfect, a fifth state, the substance of the heavenly bodies known as the quintessence.

If earth, water, air, and fire match up with solid, liquid, gas, and plasma, the quintessence corresponds with a hypothetical fifth state of matter that was dreamed up by Albert Einstein in the 1920s, long before it was first produced in the laboratory. Einstein was inspired by a young Indian physicist called Satyendra Bose, who devised a new way to describe light as if the photons that made it up were a gas. A gas was, after all, a collection of quantum particles whose behavior was already well understood. Einstein helped Bose firm up the math, but was also inspired by the young Indian's concept to imagine a fifth state of matter. Einstein believed, by applying intense cold or huge pressure to a material, it would eventually reach a state where it would no longer be an ordinary substance; instead it would share some of the characteristics of light itself. Such a state of matter is now called a Bose-Einstein condensate.

Nearly eighty years after the theory was developed, a Danish scientist working in America has used a Bose-Einstein condensate to do amazing things to light. Her name is Lene Vestergaad Hau. In 1998, Hau's team set up an experiment in which two lasers were blasted through the center of a vessel containing sodium atoms that had been cooled to form a Bose-Einstein condensate. Normally such a condensate would be totally opaque, but the first "coupling" laser creates a sort of ladder through the condensate that the second light beam can claw its way along—at vastly reduced speeds.

In the process, the photons of the second light beam—the "signal"—become entangled with the atoms in the condensate. As a long pulse of light flows into the condensate, the front part of the pulse is slowed down by the entanglement, while the rear end plows in at full speed. The result is that the light pulse is hugely compacted. A pulse that is perhaps 2 kilometers (1¼ miles) long (still a very short pulse in time, bearing in mind that light travels 300,000 kilometers in a second) can be compressed to a few thousandths of a meter.

In the first successful experiments, light was measured in the condensate traveling at around 17 meters (20 yards) per second—20 million times slower than normal. Within a year, Hau and her team, working at Edwin Land's Rowland Institute for Science, at Harvard University, had pushed down the speed to below a meter per second. A Bose-Einstein condensate isn't normal matter anyway, and this entangled mix of light and matter really is something in between insubstantial light and substantial matter. This strange mix, neither one thing nor the other, is given the romantic sounding name of a "dark state."

Producing a dark state is delicate work. To get into Hau's lab you have to take off your shoes and generally be checked out for dust, lest you contaminate the air and upset the precision optical systems. There's even a plastic curtain around the table on which the experiment is based, largely to stop interference from passing onlookers. According to Hau, this was added after a German TV crew, visiting the lab, set up a smoke generator near the experiment when no one was looking. The shamefaced journalists had intended to make the experiment's lasers visible to increase the visual impact of what otherwise was just a dull-looking piece of equipment—instead they succeeded in temporarily disabling the experiment.

But slowing down light wasn't enough for Hau's team. If the power of the coupling laser is gradually reduced, it's almost as if the process compensates for the lack of input by drawing more power out of the entangled dark state, slowing the light even further. Continue to gradually turn down the coupling laser until it is finally extinguished while the dark state is still in existence, and the light comes to a complete standstill, trapped in its entangled form. (The Uncertainty Principle is not violated, since we don't know where in the dark state an individual photon is.) Reinstate the coupling laser and the light starts moving again. In principle, Hau's team has developed a photon store, where light can be held indefinitely until required.

Even if photons can be tamed this way, there is also the problem of getting them to interact with each other. This may also be helped by the dark state. When a photon entangles with an atom in the dark state, it shares its quantum information with the atom. It is possible to effectively transfer the qubit information from a photon to the condensate and back out again.

This isn't trivial at the moment, but an experiment has shown how it might be done. Mikhail Lukin and colleagues at Harvard used a more complex version of the Hau setup. Here there was more than one coupling laser, and sandwiched between bands of dark state was what should have been a region where light traveled normally. But because it was fenced in by dark-state bands, where the light speed was zero, this "normal" light couldn't get out and was trapped, still moving, but as if it were inside a perfect mirrored box.

This is where things get a little complicated. That trapped, moving light interacts with the atoms around it. Send yet another light beam into the dark-state region bounding it and the result is an interaction between the trapped light and the new photons. In effect, clumsy though it is, it's a way of getting one photon to interact with another. Although the Bose-Einstein condensate is an intermediary, the effect is that the photons interact directly with each other. Here, hopefully, is a mechanism for light-based qubits to undergo a computational change, though at the moment it is a very complex and delicate mechanism. It's not so much data processing as light processing.

As yet the whole dark state manipulation of photons could not be conceived of as a possibility for commercial computing—it is just too fragile. But there is hope for the future. Traditionally, Bose-Einstein condensates have required heavy equipment just to keep the condensate in place, but devices based on the usually ignored magnetic effect above the surface of a microchip can hold a layer of condensate in place with no more equipment or power than is found in a laptop computer. And experiments in 2002 by Phil Hemmer of the Air Force Research Laboratory have produced similar effects to Hau's but using solid yttrium crystals, which would provide a much more practical solution to the problem.

Before the dark state was available, the best way to get photons to influence one another was using special optical materials called nonlinear media—and these still provide a strong alternative. These nonlinear devices depend on the Kerr effect. In some substances, a photon traveling through the material changes the refractive index of the medium around it, and so will influence another photon that follows in its wake. There are plenty of substances that behave like this—even ordinary glass or a bowl of water with sugar dissolved in it has a faint Kerr effect, but there is a downside when compared with a dark state: All Kerr media are highly inefficient. About the best result that can be managed is for fifty photons to be absorbed in the medium for every one that gets through, which makes this a difficult control mechanism to use in a computer that is expected to produce consistent, useful results.

If getting photons to interact with one another is a problem, getting photons and atoms to interact is relatively easy. Not only is it the basis for everything from our eyes working to the structure of matter, the interactions are well understood and accurately described by quantum electrodynamics. One way around the photon interaction problem, not unlike the dark-state trap but more simple physically, is to combine an optical cavity—at its simplest a pair of parallel partly silvered mirrors which a photon can pass into, bounce back and forth in for a while, then escape from—with a single atom, which can act as a sort of QED catalyst, enabling two photons to interact. This is still not a perfect solution—these cavities may absorb photons, and with the multiple cavities needed to build up a computer's structure, this would become a major factor—but it has some advantages over a pure photon approach.

If photons prove too slippery to handle, the qubit could be based on a more "tame" particle like an atom or an ion. In 2003, Yasunobu Nakamura and a team from Japanese electronics giant NEC successfully entangled two qubits using ion traps—devices that use powerful electrical fields to pin an ion in place in a vacuum chamber, where laser pulses can be used to change the ion's state.

Another atomic possibility is nuclear magnetic resonance, a technique that is well established in chemistry and medicine, so is well supported commercially. NMR is the effect behind the MRI scanner used to take sectional scans through human organs—the word "nuclear" was thought to be too unnerving for patients, so "magnetic resonance imaging" is the term that's used to make the medical equipment sound less scary. Not only does the wide commercial application make the technology well understood, it is likely to be a cheaper option than some of the specialist techniques if the hardware is already in production.

Nuclear magnetic resonance relies on measuring the magnetic spin effect of the nucleus of atoms or molecules—but the effect is so weak that it can only be detected by using groups of billions of atoms. Taking an approach that uses an "ensemble" of many atoms is not infeasible, but makes it much harder to distinguish the different components of the computer—as the number of qubits increases, the signal strength that has to be detected decreases exponentially. There is some hope, though, from work in 2004 at the universities of Oxford and York and of Montreal, using only a pair of nuclei to demonstrate nuclear magnetic quantum computing, and from a paper published in April 2005 by a Japanese team working with a basic but functional self-contained NMR semiconductor device. Even so, these remain early days.

With other possible qubit candidates including electrical charge and magnetic flux, it is clear that a huge effort is going into finding a working solution to quantum computation. A final option being studied with particular enthusiasm is quantum dot technology. Quantum dots are appealing because they are an extension of conventional solid-state circuitry, though the construction of them is so fine that rather than using the usual photographic processes to etch a semiconductor, circuits are sprayed onto a surface a single molecular layer at a time.

The dots themselves are tiny blobs of semiconductor that act as if they were a quantum pit in which an individual electron can be held and manipulated. Like every other potential environment for quantum computing currently being considered, quantum dots have a drawback. They can only sustain a quantum state for a very short time—perhaps a millionth of a second—before the state collapses. On the plus side, though, being solid state, quantum dots are very simple to construct and to use compared with many of the other technologies. It would be much easier to construct an array of hundreds or even thousands of qubits this way than with any other current approach. In 2004, for the first time, teams in Germany and Japan set up an entanglement between a photon and a quantum dot, raising hopes for using this technique to build solid state quantum computers.

The stark truth all researchers into quantum computers face is that it isn't enough for their equipment to operate falteringly in ultradelicate laboratory conditions. This technology may never be robust enough to appear on the desktop, but it has to, at least, be as stable as today's pampered supercomputers, and to be capable of withstanding a disastrous collapse into the state known as quantum decoherence in the timescale of the operations required.

Decoherence is the natural tendency of qubits (or anything else in a quantum state) to interact with the other quantum objects around them and to lose their unique state. For convenience, we tend to think of an experiment (or, one day, a quantum computer) as being quite separate from its surroundings. In reality, everything from the workbench to the physicist running the experiment is made up of a collection of quantum objects, and it is almost impossible in the long term to stop interactions between those particles and the ones in the experiment from destroying the purity of the quantum state holding a qubit.

Some of the early attempts at constructing qubits only remained usable for a tiny fraction of a second, and keeping qubits in one piece for minutes at a time that may be required for a calculation proves highly challenging with some of the potential quantum computer technologies. We are used to computers that have mechanisms for keeping information almost indefinitely. Conventional computer chips hold on to the information they are processing as long as they are powered up (unless they are interfered with by electromagnetic energy, such as power spikes or powerful radiation). A quantum computer may have to take a more "hot potato" approach, rapidly throwing the qubits from one place to another to keep the computer alive.

Ironically, the main weapon against decoherence is entanglement. By storing information in an entangled collection of qubits, even if the some of those qubits become corrupted by the apparatus or other aspects of the outside world, the information will not be lost. The good news has to be tempered. To protect quantum information requires more actual qubits per qubit of information used. And the more qubits there are, the harder it is to put the quantum computer together.

This is a reflection of the other big practical problem with quantum computers, scalability. A handful of qubits isn't enough to make a computer that will perform a useful task. Compared to a conventional computer, there's no need to have vast quantities of qubits—somewhere between a hundred and two thousand would be plenty for most needs—but, as more are added to the computer, there is increasing danger of rapid decoherence, extra errors are introduced, and the system becomes impractical with worrying speed.

Luckily, there are ways for quantum computer scientists to defend their hypothetical quantum computers from errors. By using several qubits to store one qubit's worth of information, a partial corruption of the data can be coped with. It's not possible to clone the state of a qubit, so you can't just keep several copies and compare them as you can with a conventional bit—but a whole field of "quantum error correction" has sprung up, providing mechanisms to process multiple qubits as if they were a single one.

It is entirely possible that none of the existing technologies will provide the "real" quantum computer breakthrough. In the development of the electronic computer, devising a better vacuum tube wasn't the answer—it took the shift to solid state for the computer to really take off. Apart from anything else, a tube-based computer with the capabilities of a modern processor would heat a good-size town. ENIAC, the first major electronic computer in America, had 18,000 tubes, was 80 feet long, and put out so much heat it required tons of cooling equipment to prevent meltdown. Imagine a vacuum tube computer equivalent to a modern 100,000,000-transistor PC. Impossible. Richard Hughes, of the U.S. National Laboratory at Los Alamos, has commented, "We're still in the vacuum tube era of quantum computing." It may well take a totally different technology to make quantum computers practical.

As we have seen, whichever technique is used to build them, quantum computers need to overcome the impossibility of copying a qubit using an entanglement-based phenomenon called "quantum teleportation." Teleportation is the strangest of all the well-established entanglement phenomena, and one that has consequences that could stretch far beyond a quicker computer, all the way into territory that has previously been the sole domain of science-fiction authors.

CHAPTER SEVEN

MIRROR, MIRROR

O'erstep not the modesty of nature; for anything so overdone is from the purpose of playing, whose end, both at the first and now, was and is, to hold, as 'twere, the mirror up to nature; to show virtue her own feature, scorn her own image, and the very age and body of the time his form and pressure.

—WILLIAM SHAKESPEARE, *Hamlet*

The quantum computers we met in the last chapter may eventually be the first capable of rivaling and even beating the human brain—a step on the dangerous road to creating conscious machines. Entanglement not only makes this old science fiction favorite a possibility, but has the potential to make another fantasy real. Teleportation.

Teleportation, matter transmission, a transporter—call it what you will, the idea is to get something solid *here* over to *there* without passing the physical object through the space in between. Somehow, we want to take a physical object and push it down a wire or broadcast it through the air. Admittedly, we can send a fax, but this is just a crude copy of the appearance of a document—it is not a true physical duplicate. The only way to make true teleportation possible is to identify exactly what makes up the original, break it down into its component parts, then, step by step, construct an identical copy of it.

That's much easier said than done. Remember the uncertainty

principle. Once we start to measure one aspect of a particle in exact detail, we become less and less certain of other measurements. We can know exactly where it is—as long as we are totally unsure of its mass and velocity. We can pin down its momentum, but only if we have no clue as to where it is. Yet to copy matter implies knowing everything about each particle in the object. This is utterly impossible.

Then there are those irritatingly vague quantum states. Measure the spin of a quantum particle and it will turn out to be up or down. But that doesn't tell you the state it was in—just the state it collapsed into. To exactly copy a particle we need to discover the probabilities that it will collapse into spin up or spin down—without ever disturbing the delicate balance of its superposed states.

And the final nail in the coffin of would-be emulators of Mr. Scott, the engineer on the starship *Enterprise*, was driven into place in the 1980s by William Wootters, of the University of Texas at Austin, and Wojciech Zurek, of California Institute of Technology. It was Wootters and Zurek who came up with the "no cloning" theorem, proving it was impossible to clone—make an exact copy—of a quantum particle. It's not that cloning a particle is difficult, not that it is beyond today's technology. Instead, it is fundamentally impossible to make an exact copy. You can never start with one particle and end with two the same.

It might seem odd that quantum objects can't be cloned. After all, we can clone a sheep, in itself a massive and complex collection of quantum objects, so why can't we clone something as simple as a photon? Unfortunately, when the "no cloning" theorem was given its name, "cloning" was used to describe a different process to the familiar biological one.

When scientists clone a sheep (or any other living thing), they are making a copy of the "recipe" that controls the complex chemicals that will build that particular creature. Even at this level, cloning will not produce an absolutely identical copy of the sheep. Small-scale mutations and the impact of the environment as the creature develops will change the outcome—which explains why identical twins, natural human clones, aren't truly identical. A biological clone is nothing more than a new batch of cookies, so to speak, cooked to a near-identical recipe.

The chemical instructions of DNA aren't capable of cloning copies of each quantum particle in the creature. From the quantum viewpoint, DNA is a lumbering great elephant of an object that has no chance of duplicating the complexity of individual quantum states. We don't need quantum-level cloning in reproduction (which is just as well, or the reproductive process would fail).

You could compare the difference between quantum cloning and biological cloning with the difference between a real-life scene that you are looking at, and a photograph of that scene. If you wanted, you could print off as many photographs of the scene as you liked and they would, effectively, be clones of one another—but you can't clone the scene itself. It is unique. However well you tried to physically reproduce the scene, it wouldn't be identical with the original, down to each blade of grass, wisp of cloud, and molecule of air. Similarly, we can "duplicate" an animal, but we can't make an exact, perfect copy of a quantum particle that includes all the detail about the state it is in.

Although its history has rarely been documented, Wootters and Zurek's no-cloning paper itself had a strange conception. A paper proposing the cloning of a photon was submitted to a scientific journal. The submission was refereed by the Italian physicist Giancarlo Ghirardi, who recommended it for rejection because, as far he was concerned, it was obvious from quantum mechanics that cloning wasn't possible. It was only when another referee recommended publication that Wootters and Zurek went into print to counter the idea, ending up "owning" an important theorem that Ghirardi had thought was trivially obvious to everyone.

However, there is a loophole in the restriction on cloning. Although you can't end up with two identical particles, Wootters and Zurek's theorem says nothing about the possibility of destroying the state of a particle while applying that state to another. That is, you might not be able to make a clone, but there is no theoretical restriction from Wootters and Zurek's work on being able to trash one particle while transferring its exact nature to another. It is this process that physicists refer to when talking about quantum teleportation.

Just because there's no absolute restriction, it doesn't mean something is possible—and the uncertainty principle and fiddly quantum states certainly make things difficult for the would-be teleporter. Yet the sort of spooky connection that entanglement provides, working at the quantum level, seems a natural way to help out with this problem, and it was exactly this circumvention that Charles Bennett, of IBM, suggested in 1993.

By a delightful symmetry, Bennett's idea was inspired by a discussion with the same William Wootters who had proved it was impossible to clone a quantum state in the first place. At a symposium arranged in Montreal by Gilles Brassard, Wootters had given a talk about the differences between making measurements on particles individually and jointly. In the discussion that followed the talk, Bennett threw in, seemingly randomly, the idea of providing one-half of an entangled pair to each of the parties attempting individual measurements to see how it would influence the outcome. Gilles Brassard later commented, "After two hours of brainstorming, the answer turned out to be teleportation. It came out completely unexpectedly."

There's a delightful contrariness about Bennett's scheme. Thanks to the uncertainty principle, the only way you can teleport a particle, the only way you can apply its state to another particle remotely, is if you can manage to do this without ever finding out what the particle is really like. In what Anton Zeilinger has called "a very elegant trick," entanglement enables us to strip the state from one particle and transfer it to another, without ever knowing what that state is. If this seems paradoxically insoluble, we should remember that entanglement is all about making paradox real.

The trick here is to use two sets of information, one transmitted by the spooky quantum link of entanglement but never observed by anyone, the other known and sent by a conventional channel—say, by radio. (The need for both sets of information keeps teleportation safely out of the faster than light tangles of chapter 5, though it is not imposed to satisfy the time COPS: the necessity falls out of the physical requirement to transfer a quantum state.) No one ever finds out what is communicated by the entanglement link, so the quantum state isn't disturbed by it.

When you get down to the detail, quantum teleportation has a very real resemblance to a magic trick—you need to keep an extra particle up your sleeve. To undertake teleportation there are not two but three particles involved in the process. We start with an entangled pair of particles, and send one of them to the receiver station. This can be done at any time, days, weeks, or decades before the teleportation process, provided the entangled particles can be kept intact in the entangled state.

With our entangled link in place, we come up with a third particle, the original: the one we want to teleport. This is made to interact with the entangled particle at the sending station, resulting in an instant (but never directly measured) change in the remote entangled particle. The transmitting station then makes some measurements of its particles and sends the details to the receiver. By now, as a result of taking those measurements, the original particle has lost the fundamental characteristics that make it the particle it was. Its quantum identity has been destroyed. If the particle were part of a larger whole, and this process was done to every particle in the object, that object would now be no more.

At the receiver end, the second entangled particle is modified by the information sent across the radio link—the information that is known. Now the receiver particle has become, to all intents and purposes, that same particle that was originally examined at the transmitter. This all sounds rather vague, but in specific circumstances the teleportation process is quite straightforward.

For example with two photons of light, if we use polarizations at right angles to represent 0 and 1 (say, 0 is horizontal, 1 is vertical), there are four possible outcomes of the sender measuring the polarizations of the pair of photons she has: 00, 01, 10, or 11. Having made that measurement (and destroyed the state of both her original photon and her entangled photon), she can send the result to the receiver. He then puts *his* entangled photon, already changed as a result of the measurement made at the other end, through one of four different processes (the simplest being "do nothing" and the others being different quantum gates like the X gate we have already met on page 189). The result is to make the second entangled photon at the receiver indistinguishable (apart from position) from the original teleported photon.

The concept was made real simultaneously in 1997 by Anton Zeilinger and his team in Vienna, and Francesco de Martini and colleagues in Rome, using an idea from Sandu Popescu. In the experiment, the polarization of one photon was transferred to another. Many physicists are extremely cautious about the way they describe their achievements, but some are showmen and can't resist a dramatic flourish—not a bad thing in a scientific world than can verge on the dull. Careful scientists will tell you that quantum teleportation has nothing to do with Star Trek–style transporters. It's just a process that involves duplicating the state of a single particle. But Zeilinger can't resist the dramatic touch, and he made it clear in his paper on the experiment that he believed that it was about something significantly more exciting:

> *The dream of teleportation is to be able to travel by simply reappearing at some distant location. An object to be teleported can be fully characterized by its properties, which in classical physics can be determined by measurement. To make a copy of that object at a distant location one does not need the original parts and pieces—all that is needed is to send the scanned information so that it can be used for reconstructing the object.*

This is, Zeilinger is saying, the real thing, even though on a very small scale.

By 2004, Zeilinger and his team had achieved teleportation over significantly greater distances—in fact, across the river Danube. A year after their groundbreaking long-range transmission of entangled photons across the Danube (see page 83), the Austrian team was back in the sewers, this time achieving teleportation from one side of the river to the other. (Quantum entanglement experimenters seem to have a functional relationship with the sewage system rivaled only by utility workers and Buffy the Vampire Slayer.)

As always with teleportation there are two "channels," one carrying the entangled particles, the other transmitting the conventional information that will be used to complete the teleportation process. Entangled photons were pumped along a fiber-optic cable running through the sewer system under the Danube, while the conventional information was beamed by microwave for 600 meters (666 yards) across the river. This may not seem groundbreaking but, as their paper in *Nature* commented, they had "demonstrated quantum teleportation over a long distance and with high fidelity under real world conditions outside a laboratory."

This is a significant blow to those critics who have said that teleportation could only occur under highly controlled laboratory conditions. The team points out that it's also possible that this technique could be used as an alternative approach to make quantum repeaters (see page 80) that would enable entanglement to be shared anywhere around the world, as teleporting an entangled particle transfers the particle's state, including its entanglement.

As this demonstrates, even if there never can be "real" teleportation of physical objects, it doesn't mean that this isn't a development of great importance. Teleportation even in its limited form will prove vitally useful in making quantum computers real. Quantum computers, as we saw in the previous chapter, rely on qubits, by means of which information is stored in the quantum state of a particle. This may be very powerful, but it is also difficult to transfer that quantum state safely from place to place within the computer—or even between two quantum computers.

Teleportation means that, provided a supply of entangled particles is available, something that is now relatively easy to achieve, a qubit can be teleported from one place to another using only a conventional link. Thus, a satellite pumping out entangled photons to two locations could not only be used to provide the key for quantum encryption, it could enable quantum computers in two locations to swap qubits over the Internet.

It might seem that it is going to be just as difficult to get the entangled "carrier" particles from A to B as it would the qubits, but the big difference is that those entangled particles are vanilla—they don't carry the information until after they are in place. So in the real world, where some particles are inevitably lost along the way, we can send as many entangled pairs as we like out, ensuring each end of the connection has a supply—once that is achieved the individual qubit, which we can't afford to lose, can be teleported.

In some ways, the nearest we've yet come to the teleportation of an object didn't involve the final teleportation part of the process itself. In late 2002, Eugene Polzik and his team at the University of Aarhus, in Denmark, managed to entangle two clouds of cesium, each containing billions of atoms (around 1,000,000,000,000 in each)—in effect an object, if a rather insubstantial one, and one big enough to be visible to the naked eye. This is a significant step forward because entanglement was, until this experiment was undertaken, largely seen as something that applied to at most a handful of quantum objects at a time.

In the Aarhus experiment, a pulse of laser light is blasted through the two samples of cesium, pushing their spins into an entangled state rather as if two spinning tops were lashed by the same whip. This is a step on the way to demonstrating partial teleportation by transferring the magnetic state of one cloud to another.

Can we ever use teleportation on solid objects with structure—perhaps even with life? Even for a single particle this is a nontrivial challenge. Teleportation experiments to date have focused on a single property of a particle—its spin, for instance—but to truly teleport a particle it would be necessary to teleport all the properties separately. As Professor Artur Ekert comments, this is "mathematically possible, but it may be way more difficult to do it in the lab!"

However, one possible opening for this was developed in 2002 by Sougato Bose, of the University of Oxford, and Dipankar Home, of the Bose Institute in Calcutta. If two particles—electrons, for instance—are sent through a beam splitter, it is possible for both to go down either path, or one to go down each. (As we've seen, a beam splitter is the device that lets some particles go one way and some another, such as a half-silvered mirror.) Bose and Home have shown mathematically that when the electron pair split, one in each direction, they should become entangled. In principle, this could be applied to any particle that can undergo quantum mechanical "superposition of states"—effectively being in two states at the same time.

What is exciting about this discovery is that it's not just the tiniest of quantum particles like photons and electrons that can be put into superposed states. Anton Zeilinger and his colleagues in Vienna have demonstrated superposition with the delightfully named "buckyballs." These are large molecules of carbon with sixty atoms, shaped like a miniature soccer ball. The name, another rare example of scientists coming up with something better than a challenging collection of syllables, is a shortening of "buckminsterfullerene," so called because the structure of the carbon molecules resembles the elegantly curved geodesic domes designed by American engineer and architect R. Buckminster Fuller.

Buckyballs don't present a maximum size limit, though. Zeilinger's team has achieved superposition of bigger molecules with a size comparable to that of living bacteria. In principle, if Bose and Home's technique works, it could be applied to entangle any particles that can be in a superposed state, up to and including Zeilinger's large molecules or bacteria.

At the time of writing, that experiment hadn't been carried out. There are some practical difficulties of keeping a bacterium alive in the conditions needed for the experiment (a vacuum, low temperatures, and so on), though Zeilinger suggests that it might be possible to genetically engineer a protective casing around a bacterium. However, the point is not so much doing the experiment as demonstrating that quantum effects can escape from the prison of the very small to influence objects we would generally regard as "real" and tangible.

There's a long way to go down a path that could see the transportation of a living creature, however simple. The process would have to begin with something like a tiny crystal, move on to a virus (not truly alive, but having more of the complexity of a living structure), and eventually work up to a bacterium, a truly living organism. And the leap from there to large-scale life, such as a human being, is even greater and probably will never be practical. It would also require more of a step-by-step approach—there is no way to get a whole person into a superposition of states—and the sheer number of molecules in the human body seems to be an insuperable limit. This is despite the fact that the urge to get bigger and bigger items into superposition has led to Anton Zeilinger being accused of wanting to drive a truck through an interferometer.

This is not the act of wanton vandalism it sounds to be. The idea is not to destroy the interferometer, a delicate piece of lab equipment (it's the device used to measure interference in the sort of superposition experiments described above), by actually crashing a truck *into* it. Instead, the suggestion is that Zeilinger wants to dispatch a truck through two different paths of the same piece of apparatus simultaneously, putting it into the superposition of states that has been achieved with buckyballs and other large molecules. An interferometer would *then* be used to detect the interference pattern between the states of a truck within that other equipment. Why such a bizarre desire?

In fact, Zeilinger initially seemed nonplussed when the whole of idea of him wanting to drive a truck through an interferometer was put to him, denying he ever said it. Where this vivid image appears to have come from is that Zeilinger has dismissed as nonsense an old argument that said that it would never be possible to see quantum interference using an object as large as a truck, because the truck is much bigger than its de Broglie wavelength, a feature of all quantum objects.

The interference of a quantum particle with itself, whether we're dealing with electrons or buckyballs, is often described using the concept of a de Broglie wavelength—the idea being that the particle is acting as if it were a wave with a particular wavelength, producing interference that resembles the interaction of ripples on a pool of water. However, Zeilinger points out, the truck argument isn't valid because the large molecules that have already demonstrated interference are themselves much bigger than their own de Broglie wavelengths. "I don't think we will see interference with trucks," commented Zeilinger drily, "but not for this reason."

Zeilinger is almost certainly right, but this doesn't absolutely dismiss the idea that it may one day be possible to produce a teleportation system that can handle a real, physical object and even a human being. It is dangerous to absolutely rule it out. After all, attempts at predicting the scientific future go wrong with notorious regularity.

Sometimes a projection of an idea into the future oversimplifies the difficulties involved in a practical implementation and brings a development forward too far in time, or even makes the impossible real. Take a look at the movie *2001: A Space Odyssey,* for example. Okay, it's fiction, but Arthur C. Clarke and Stanley Kubrick did their best to make the science in the movie a serious projection of what was believed possible at the time. Released in 1968, we have to remember that *2001* was peering more than thirty years into the future, a comparable leap forward to the gap between the Wall Street crash of 1929 and the swinging sixties.

Some of the errors in *2001* were caused by the ordinary random fluctuation of events that simply could not be predicted. For example, the shuttle rocket used to reach the space station was operated by the long-disappeared airline Pan Am (Pan American World Airways). At the time the movie was made, Pan Am was one of the biggest names in American aviation, a warmly familiar company that no one could have envisaged collapsing as it did in 1991. But other innovations, more directly bedded in the prediction of the development of technology, were wildly optimistic.

This is obvious in the existence of the commercial space shuttle, complete with flight attendants, or in the HAL 9000 computer that appeared to have consciousness and independent thought, but was more subtly and drastically brought out in the pay phones. These had large-screen, full-motion video. Yet several years after that film was set, this is still a long way off being widespread reality (and we are more likely to make a video call on the Internet than through a traditional telecom connection).

When UK science writer John Gribbin was writing about the very early days of quantum teleportation in 1995, he described Bennett's 1993 paper and discussed the possibility of teleportation in positive but hardly hurried tones. "Given the ingenuity of the experimenters . . . there must be a good chance that before 40 more years have passed they will be sending electrons from one side of the lab to the other, or even around the world . . ." In fact a photon was teleported just two years later, and an atom (or more precisely an ion) in less than ten years. Gribbin's prediction came true much sooner than expected.

It seems reasonable right now to say that teleportation of something as complex as a virus might be achieved within twenty years. Anton Zeilinger is doubtful about going much further. "Nothing in principle limits [the size of the object that can be teleported]. But for sufficiently large objects—probably anything living—teleportation is still a fantasy, but you never know!" As Zeilinger, himself, has commented, "An experimentalist should never use the word 'never.' Some of the experiments we are doing today I would never have believed possible ten years ago."

Just imagine for a moment that it were possible to send a human being through a transporter based on quantum entanglement. In many ways, the concept is attractive. In principle, you could cross the world at the speed of light, though in practice there might be a few minutes delay to handle the initial scanning and the reestablishment of all those quantum states. Yet would anyone be prepared to undergo the risks involved just to get somewhere more quickly?

To be teleported, every atom in your body would have to lose its quantum uniqueness. It would involve nothing less than total disintegration. Yes, the outcome would be a perfect copy with all your memories and personality, but would it be you? If you believe in the existence of a soul, would you expect your soul to somehow migrate to the new body? If, as with many scientists, you thought that your mind was just a function of the physical body, then would it be enough for you that an identical copy of your mind existed? What about the "you," the consciousness that makes you what you are?

The decision is more difficult that it seems. After all, our bodies are changing all the time, replacing molecules on a daily basis. In a very real sense, you aren't the same "you" you were ten years ago. And every night we experience the loss of consciousness, detaching ourselves from reality until waking up. Is this so very different?

Certainly, people being what they are, someone could be found to volunteer to undergo teleportation. Others might find that they had little choice. It's easy to imagine military personnel being teleported to remote locations with little concern for their personal feelings. Were it possible, someone would pass through that system and come out the other side. And to everyone watching, it would be the same person who arrived at the distant destination. Not similar, but the same. Even so there would be plenty of others for whom teleportation would never be acceptable, and I have to confess, I would be one of them.

At first sight, teleportation is the most extreme of the potential applications of entanglement—but it could be just the beginning. Some scientists have linked entanglement to telepathy, to the source of a particle's mass, and even to life itself.

CHAPTER EIGHT

CURIOUSER AND CURIOUSER

Prais'd be the fathomless universe,
For life and joy, and for objects and knowledge curious,

—**WALT WHITMAN**, *"A Backward Glance O'er Travell'd Roads"*

What we have seen is only the start. Entanglement lies at the heart of not one, but a collection of major breakthroughs. As we build on this science at a tipping point, it is very likely that even more remarkable applications will emerge. This is despite the fact that a minority of scientists are not happy, even at this stage, with the whole basis for entanglement, effectively denying the existence of what is by now an indubitably well documented effect.

These doubts are nothing new. As we have seen, Einstein was uncomfortable with quantum theory, and over the years many others have felt discomfort at the disparity between the world as we see it—solid, reliable, full of identifiable objects that stay where you put them—and the accepted model of the quantum world where nothing is really what it seems. The mainstream approach is almost to pretend this division doesn't exist, to take the attitude "There's nothing we can do about it, so we might as well ignore it," but some theoreticians continue to worry away at the division between quantum theory and the observed world.

This discomfort has led to a number of alternative explanations of quantum phenomena, most notably in ways of explaining the strange duality by which quantum objects seem sometimes to be particles and sometimes waves, which suggests they are truly both at the same time, or in theories that explain the observed quantum peculiarities by theorizing that the world splits into two each time an event occurs, creating the sort of multiple parallel universes as described in chapter 2.

These different interpretations of quantum theory are widely discussed in books on quantum mechanics, and don't need to be covered in any depth here as they have limited relevance to entanglement—in the end, we know entanglement works experimentally—but it is important to address the concern felt about the lack of substance, almost the lack of reality, in the quantum world that continues to cause unease even to experts in the field, such as John Bell.

A simplified picture of the real concern many feel is to ask, "How can the components that make up the world be vague things, particles at some times, waves at others; particles or waves that don't even have a clear idea of location? How can solid, substantial objects like my desk in fact be a collection of entities that are so fuzzy in nature that they could (with varying probabilities) be anywhere in the universe?" I think it's useful when dealing with this problem to remind ourselves—as we did in chapter 2 with Shrek and the onion—that science, and physics in particular, is about building models, not about defining the nature of absolute reality.

We can never directly examine a photon or an electron. We can't handle them, touch them, look at them, or taste them. Although a photon can trigger our optic nerve, resulting in sight, we don't "see" photons, but rather our eyes respond to the energy that they carry from somewhere else. We can't pull an atom apart by hand and see how it works as we might disassemble a clock. All we can do is build a mental model, a physical metaphor for the quantum object, and see how well the model matches up to reality. We ask what the model would do in various circumstances, and compare its reactions with the real world.

This importance of models to physics is illustrated well by an old joke that is usually an excellent way to divide scientists from others in a crowd (the scientists are the ones who laugh at it; the rest just look baffled, or smile politely).

> *Three people—a geneticist, a dietician, and a physicist—are arguing about the best way to produce a winning racehorse. The geneticist says, "It's no problem if you follow good genetic principles. Just breed from winners, selecting for the winning characteristics, and in a number of generations you should have yourself a winning racehorse." The dietician says, "No, you're wrong. I'm not denying the importance of genetics, but to ensure a winner, we've got to give the horse a planned diet combined with appropriate exercise that will ensure optimum performance." The physicist shakes her head sadly. "Look," she says, "let's imagine that the racehorse is a sphere . . ."*

The point of this joke is that physics often works by making vast simplifications, and by using something we *do* know about, in the joke a sphere, as a model to stand in the place of something we know less about—in this case racehorses. When we describe light or an electron as a wave or a particle, what we really mean is that we are using the model of waves (like the actual, real ripples we see on the sea) or the model of particles (like a stream of very, very tiny bits of dust). But what shouldn't be implied is that light or an electron *is* a wave or that it *is* a particle. Attractive though it is to have something that we are familiar with to hang on to, we always have to remember that we are dealing with a model, not the real thing.

What is light or an electron, then? Light is light. An electron is an electron. They happen to exhibit strong similarities to waves and particles some of the time, which can be very useful when we are trying to predict what they will do, but that isn't what they are. If you can take the step of seeing this, the whole concern about quantum theory is better put in proportion. The only reason the behavior of a photon appearing to pass through two slits at once, or entanglement working at any distance, is odd, is that we are letting the models take over, giving them more weight than they deserve.

This doesn't mean that anything goes. Our models are very useful, and we can continue to get better and better predictions of what the world can do, of what it is *like* (as opposed to what it *is*), but we shouldn't expect absolute understanding ever to come from working with a model.

This generates a trap for would-be speculators. The fact that we're dealing with models of reality makes it easy to come up with ideas that don't have any direct basis in experiment. For example, entanglement *could* be responsible for all kinds of phenomena. Some have argued, for instance, that consciousness is a quantum phenomenon involving entanglement, but the biological evidence for the mechanisms of the brain suggest that there is no need to resort to such complex and delicate phenomena to explain what appears to be a very robust and invariant capability.

This doesn't make all speculation worthless, though. Not only can it produce new and wonderful ideas—arguably all modern physics originates from a handful of vibrant speculations that challenged traditional science at the start of the twentieth century—but it can result in a very healthy shaking up of what can otherwise be entrenched and self-satisfied thinking. One piece of speculation about a possible natural application of entanglement that was bound to put the cat among the pigeons came in 2001 from Nobel Prize–winning physicist Professor Brian Josephson.

Josephson has become a pariah in the world of science. He is tolerated but rarely shown the respect that might be expected for a Nobel Prize winner. Josephson was always unconventional—his initial inspiration that led to his conceiving the Josephson Junction, the quantum device that earned him his Nobel Prize, came when, as only a graduate student, he was prepared to stand up at a conference and challenge the results being put forward by a senior, highly experienced physicist (and a Nobel laureate to boot).

More recently Josephson has headed up the Mind Matter Unification Project, part of the Theory of Condensed Matter Group, at Cambridge. In his later career, Josephson has committed what many scientists regard as a major transgression—he has displayed remarkable openmindedness, which is regarded by many as little more than gullibility.

There is a fundamental dichotomy when trying to advance scientific knowledge. If you have a totally open mind and test all possibilities, you can become trapped forever on wild-goose chases, never pursuing a useful lead. But the other extreme is just as bad. It's all too easy to ignore anything that goes against current thinking, which is an inevitable recipe for stagnation. Such an attitude would have left us little advanced from the authority-obsessed years, when it was not considered appropriate to argue with an accepted viewpoint. This was the ethos that kept science frozen with little development from the time of ancient Greeks all the way through to the scientific renaissance, when it was swept away by the likes of Galileo and Newton. We should have learned by now that it can only pay to question received wisdom.

Josephson displays some of the best and, to be honest, the worst attributes of the open minded. He is prepared to look into very different ways of thinking about nature and the universe—the kind of approach that led Newton to his novel views on optics, or Einstein to relativity. But at the same time, his support for generally dismissed concepts, such as the memory of water as an explanation for the apparent effects of homeopathic medicine, has led to an unfortunate tendency for the scientific community to dismiss anything he says. Josephson may not always be right, but then neither was Newton, Einstein, or Feynman. In truth, the attitude of his peers has been appalling, implying that certain fields Josephson studies are simply not worth investigating, and can even become tainted by an interest from Josephson.

A hallmark of Josephson's work is his enthusiasm for the unexpected. A number of books quote him as saying that Bell's inequality is "the most important recent advance in physics." When I asked him why he had said this, he told me that it was because "it showed the strangeness of nature in an explicit way. You can't have a causal model without connection at a distance." Strangeness is something that Josephson clearly relishes.

Brian Josephson has come up with many controversial ideas that don't rock the boat outside the small world of physics, but one of his public statements caused a brief uproar that escaped into the wider world. The speculation on entanglement—and there was no question from anyone, Josephson included, that this was anything other than speculation—came in the unlikely form of a booklet published to accompany a series of commemorative stamps.

On October 2, 2001, the UK's Royal Mail launched a set of stamps to celebrate the hundredth anniversary of the Nobel Prize. There were six stamps, one corresponding to each of the prize categories. For the accompanying presentation pack, the Royal Mail commissioned a "Nobel reflection" from six relevant Nobel laureates. Professor Sir Harold Kroto covered chemistry, Professor James Mirrlees wrote on the economic sciences, Sir James Black weighed in on medicine, Sir Joseph Rotblat held forth on the subject of peace, Seamus Heaney wrote on literature, and Professor Brian Josephson contributed the physics entry.

Each of the stamps had some unique feature. The physics stamp itself featured a hologram of a boron molecule; the peace stamp was embossed, providing a unique texture; physiology and medicine was represented by a scratch-and-sniff stamp smelling of eucalyptus; while the literature stamp had T. S. Eliot's "The Addressing of Cats" microprinted on it; economic sciences was represented by a stamp featuring the indented lines of intaglio; and the chemistry stamp had an image in thermochromic ink that only revealed its contents when warmed. Josephson, however, given a free rein as to the subject of his reflection, decided to describe entanglement, and to include a dramatic piece of speculation in his very short text.

> *Physicists attempt to reduce the complexity of nature to a single unifying theory, of which the most successful and universal, the quantum theory, has been associated with several Nobel prizes, for example those to Dirac and Heisenberg. Max Planck's original attempts a hundred years ago to explain the precise amount of energy radiated by hot bodies began a process of capturing in mathematical form a mysterious, elusive world containing 'spooky interactions at a distance', real enough however to lead to inventions such as the laser and transistor.*
>
> *Quantum theory is now being fruitfully combined with theories of information and computation. These developments may lead to an explanation of processes still not understood within conventional science such as telepathy, an area where Britain is at the forefront of research.*

In that final sentence, Josephson threw down the gauntlet to those who despised an open mind. In effect, he was suggesting that, should telepathy exist, one possible mechanism for it to work was quantum entanglement. As there is no conclusive experimental evidence for the existence of telepathy, Josephson was being intentionally provocative, and his critics were not slow to respond. In a news item in the September 27, 2001, issue of the prestigious journal *Nature,* headlined "Stamp Booklet Has Physicists Licked," Erica Klarreich described a backlash that was real, despite typical British restraint.

She commented that Robert Evans, a physicist at the University of Bristol, said that he was "very uneasy" about a publication from the Royal Mail declaring quantum physics had something to do with telepathy. Karreich also quotes Kathryn Hollingsworth, a spokesperson for the Royal Mail, saying, "If it transpires that what he's suggesting doesn't have a scientific basis, perhaps we should have checked that, but if he has won a Nobel Prize for his work, that should give him some credibility."

Soon after, *The Observer,* one of Great Britain's most respected Sunday newspapers and anything but a tabloid scandal sheet, pushed up the temperature of the debate. Science editor Robin McKie told readers, "Scientists are furious," and quoted David Deutsch, of Oxford University and a longtime opponent of Josephson's, as saying, "It is utter rubbish . . . The Royal Mail has let itself be hoodwinked into supporting ideas that are complete nonsense."

Josephson struck back in a letter to *The Observer* and an interview on the UK's leading current affairs national radio show, *Today,* better known for grilling politicians than exploring the complexities of science. Interestingly, the arguments presented were not really about whether quantum entanglement could provide a *mechanism* for telepathy, but were stuck in the entrenched battlefield over telepathy's existence. The counterargument could be summarized as, "Telepathy doesn't exist, so there is no need to explain it," while Josephson cited various possible experimental evidence for telepathy in his response.

Is Josephson barking up the wrong tree? As telepathy is still more a subject for mystical mumbo-jumbo than hard scientific investigation, it is hard to say. Yet Josephson's speculation was not based solely on the thought that telepathy could be linked to quantum theory "because that's all very fuzzy anyway," as was claimed on *Today* by the debunker of all things paranormal and professional skeptic "The Great Randi," who is hardly overqualified to speak on quantum theory. Josephson's comment was based on a speculative theory of the mind, which may or may not be true, but did provide a sensible link to his conjecture.

When I recently met up with Josephson in Cambridge, snatching a brief version of a traditional academic pastime, partaking of afternoon tea, in the incongruously ultramodern setting of the university's Department of Mathematics and Theoretical Physics rather than an ancient paneled common room, I asked Josephson whether his intent in writing the article to accompany the stamp had been provocative or serious. He paused for a moment and then quietly but firmly said "both."

Brian Josephson feels there is an opportunity for a synthesis between biology and quantum mechanics. Entanglement, he points out, has shown that different components of a system can be connected remotely—if telepathy were shown to exist, it could well involve parts of the biological systems of the brain having a similar remote connection. "A real understanding of quantum mechanics eludes us: we tend to have too local a view of nature."

Though the existence of telepathy remains highly contentious, physicists have come up with a thought experiment that demonstrates what Gilles Brassard has called pseudo-telepathy. As we saw in chapter 5, entanglement can't be used to send a message at faster than light speed, but an entangled link can make it possible to win in a game that would otherwise require communication to succeed. Like much of game theory, the mathematics behind gaming, this has potential real applications in interactions between human beings, all in some sense a form of game—but let's see first how entanglement helps the players.

Game theory may sound like the methodology of winning at Monopoly or triumphing at Half Life, but as usual science works on a simplified version of reality. Much of game theory depends on the actions taken by two or more individuals presented with simple choices. Often the players have limited ability to communicate. Depending on the choices they individually make, they may collectively win or lose. It's often about studying the strategy for a single move rather than the game as a whole.

Perhaps the best known example in the lexicon of game theory is the "prisoners' dilemma," which is used as a model of cooperation, a model that lay behind much of the thinking in the cold war. It's a sobering thought that the advisors of generals and presidents were pointing out the mathematics of civilization's survival based on this very simple game.

In the prisoners' dilemma, the simplest scenario is that there are two prisoners, both locked in separate cells, unable to communicate with each other. Each of the players has the choice to cooperate or to defect and betray the other prisoner. If both cooperate, each gets a positive result. If one cooperates and one defects, the defector gets a bigger benefit than he or she would from mutual cooperation, and the cooperator re-

Figure 8.1. The prisoners' dilemma.

	Second prisoner cooperates	*Second prisoner defects*
First prisoner cooperates	Both benefit somewhat	First suffers badly, second benefits highly
First prisoner defects	First benefits highly, second suffers badly	Both suffer somewhat

ceives a strong negative result. If both players defect, the outcome is a weaker negative outcome for each.

On the assumption that the other person *may* defect, the "rational" action for a player seems to be to defect. After all, by defecting, a player ensures that the worst he or she will suffer is a mild punishment if the other player also defects. Should the other player cooperate, the result will be a positive reward. Yet a lone defecting player's reward only comes at the expense of the other player's suffering. The win-win situation where both cooperate is both morally preferable—everyone benefits—and better overall in the long term of a series of games. After all, most interactions between two people involve not one but a series of "games." If one player defects all the time, the other is likely to retaliate—and this will end up over the series in both players being losers.

The cold war application of the prisoners' dilemma involved preemptive nuclear strikes. Some argued that a big enough nuclear strike would be a one-off play—there would be no series of events, as the enemy would no longer be able to take part in the "game." This resulted in the chilling advice that the rational approach was to get in first and destroy the enemy. Thankfully, common humanity inclined the "players"—the Americans and the Russians—to adopt the win-win strategy.

Part of the point of game theory is to look for strategies that will best handle any particular situation (this is, after all, only a game in name)—and it is here that entanglement can lend a hand. Let's imagine a very simple game: We take two individuals and stop them from communicating. This "stop them from communicating" requirement is absolute. We don't trust our players—we have to assume they will cheat if they are able to. It's not enough to search them, or shield the locations from radio communication. These are clever folks. We have to assume that if it's physically *possible* to cheat, they will find a way.

Luckily, Einstein has provided us with a surefire defense. Let's leave one player on the Earth and send the other one off to Mars (this is only a thought experiment—it's not necessary to do this in practice). Now, it takes four minutes at their average distance apart for a beam of light to get from one planet to the other. That beam of light is the fastest thing possible. We've shown that, even with an entangled link, it is impossible to send a message faster. So all we've got to do is to insist that each player comes up with a response in less than four minutes of hearing the choice they have to make (the choices are announced simultaneously). That way, even if they did communicate, they would have to make the choice before the communication could have gone from one to the other. It's foolproof—and technology proof. The players can't cheat by secretly cooperating in any way.

However, we also have to make sure the game itself can't be beaten by applying a clever, prearranged strategy with no subsequent communication. Simple games are often easy to overcome this way, a technique that is often used in stage acts that appear to feature telepathy. Here's a simple demonstration of apparent telepathy that can be used to fool the vast majority of people: One of the telepathic pair is sent out of a room. While she is out, the audience in the room selects an object. The outsider then comes back in, and the "broadcasting" telepathist starts to point at objects around the room. His confederate says "no" each time, until he points at the right object, when she says "yes."

The amazing thing is you can repeat the exercise as many times as you like for any object, and the audience can take whatever precautions they like (apart from blindfolding the broadcaster). The trick works without any direct communication, relying purely on a strategy. In this case, the two telepathists agree that the object before the right one will be a particular color. Blue, say. The broadcaster goes around the room, pointing at nonblue objects. The penultimate object pointed to is blue. Then comes the actual selection. Lo and behold, amazingly, the confederate says "yes." This is the sort of strategy for cheating that has to be avoided.

Let's go back to our two players, one on Earth and one on Mars, and get them playing what must be about the simplest game imaginable. Each player is asked to say either yes or no. If their results are the same, the players lose. If they are different, the players win. If each chooses the word they say at random, we'd expect it to be a little like a coin toss. Half the time they would come out with the same result and lose. Half the time they would differ and win. (That's on average, of course. Any particular round would be a definite win or lose.) But with a simple strategy, they can win all the time.

Simplest of all, the Earth player could always say yes and the Mars player could always say no. Every time a winner, and no communication required throughout the game. Or to make it more interesting, on every odd play Earth could say yes, on every even play no, with Mars reversing this. And you can imagine taking your strategy to more and more levels of strategic rules to make the guess appear random without its actually being so.

That's trivial. "So what?" you probably say. The reason these games are interesting is that they can be taken to another level of complexity and still, apparently psychically, be beaten if quantum entanglement can play a part. Perhaps the best example of such an experiment was suggested by Padmanabhan Aravind, of Worcester Polytechnic Institute, in Massachusetts. His game involved a three-by-three square grid in which the two players had to put 0s or 1s in a row or column.

The rules of the game seem odd, but are necessary to make the exam-

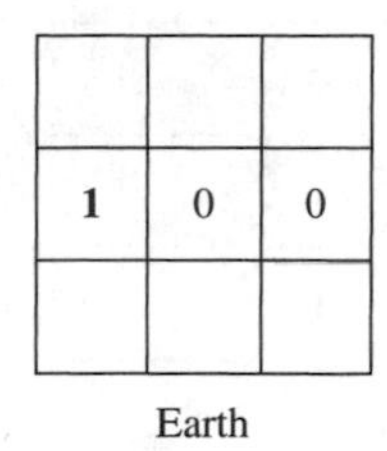

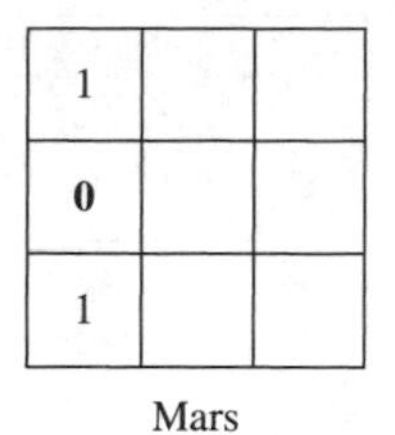

Figure 8.2. A losing play—the intersecting square is 1 on Earth but 0 on Mars.

ple work. One player deals with the rows of the "magic square," the other with the columns. When a round is started, Earth (say) is given row 2, while Mars is given column 1. Before the four-minute communication barrier is breached, Earth has to fill in her row, while Mars fills in his column. They can put 0 or 1 in each square, but Earth's must add up to an odd number, and Mars's must add up to an even. For the players to win, the square that both of the players fills in, the square that forms the overlap of their respective row and column, must have the same value.

The difference between this game and the yes/no game is that it isn't possible to agree on a strategy in advance that will result in winning the game more often than is dictated by the outcome of random chance, because the distant player has no idea which row or column has been chosen on the other planet.

Here's the clever bit. The game's organizers allow the participants to carry a stock of entangled particles. After all, we know that it's not possible to send information across an entangled link. But when the row and column numbers are announced, each player takes a pair of particles and puts them through a quantum gate that corresponds to the row or column that has been chosen. After the transformation by the appropriate gate, the players can read out a pair of values. Earth might get 1 0, and Mars 0 1. These values are used for the first two numbers in the

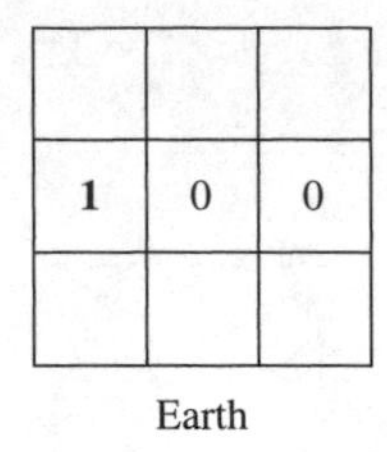

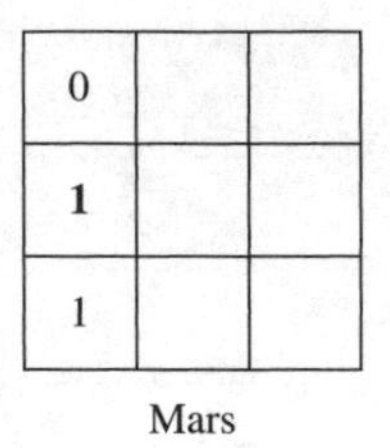

Figure 8.3. A winning play—the intersecting square is 1 on both Earth and Mars.

square and the third is deduced from the odd/even rule. The outcome is random—there is no message passed from one to the other—but because of the spooky entanglement connection, the output guarantees that the squares chosen will win.

There still isn't a way to send a message, but this application of entanglement comes the closest yet to sharing true information only across a quantum link. Of itself, the game isn't going to change the world; nonetheless it's easy to see why Brassard referred to this as pseudo-telepathy. It really is as if there was some mental connection between the two individuals, because no message can pass from player to player in the time it takes to make the play.

Although this game may not have any real world applications, there are others that correspond to business and political negotiations, just as the prisoners' dilemma game was a crude model of a nuclear standoff. One interesting possibility is that systems derived from the "magic square" entangled game may facilitate fairer outcomes of negotiations with hidden inputs, by which means some contributors can freeload on the generosity of others. Either way, it's a remarkable demonstration of just how far entanglement pushes the boundary between the real and the quantum world.

Pseudo-telepathy doesn't excite the same sort of negative reaction as some other speculation about entanglement, because it's wrapped up in the eminently mathematical logic of game theory—it is possible that part of the negative reaction to Josephson's suggestion with respect to *true* telepathy was due to the unfortunate degree to which quantum theory has attracted attention from nonscientists who are excited by an apparent similarity between the strangeness of the quantum world and mystical ideas from Eastern philosophy. It is unfortunate, though inevitable, that quantum theory has led to more philosophical discussion than has any other aspect of science.

After all, as quantum entanglement demonstrates, the quantum world is not bounded in the same way that we expect the "normal" world to be. And the sheer fact that the act of measurement changes the quantum system brings to quantum science a human flavor that gives it a particular appeal. Quantum science has to accept that the experimenter exists and is part of the experiment, an acknowledgement that appeals to the postmodernist urge to deny the existence of objectivity. While everyone agrees that quantum theory as it stands is very good at predicting the phenomena we experience, there has always been argument over the interpretation of it, and this "fluffiness" appeals to some who argue that quantum physics ties in well with traditional Eastern philosophy and religions.

Michael Shermer, *Scientific American*'s resident skeptic, points out the ease with which "New Age scientists" can pick up a little quantum jargon to spice up their ideas with a little real science veracity. Shermer picks out as a great example of what the physicist Murray Gell-Mann described as "quantum flapdoodle," a film called *What the #$*! Do We Know*. Shermer quotes a scientist in the film as saying, "The material world around us is nothing but possible movements of consciousness. I am choosing moment by moment my experience. Heisenberg said atoms are not things, only tendencies." Shermer challenges the quoted scientist to jump off a twenty-story building and test the tendency of the ground to flatten him.

Perhaps the strongest influence on the idea that quantum theory is somehow a vindication of ancient Eastern wisdom was a book called *The Dancing Wu Li Masters,* written by Gary Zukov in 1979 and still selling today despite being wildly out-of-date. Although supposedly a description of modern physics and not an attempt to integrate it with Eastern philosophy, Zukov goes to great lengths to bring out the aspects of quantum physics that would appeal to a navel-gazing generation and that reinforce the idea that there is no objectivity, no truth beyond reality as we perceive it to be. Zukov's portrayal of the mystical nature of quantum physics and the philosophical concerns of quantum physicists was wide of the mark. As John Bell commented,

> *[Gary Zukov] gives the wrong impression of what is happening in physics institutes. People are not all desperately discussing Buddhism and Bell's theorem and the like. A precious few are doing that, whereas his book gives the impression that that's what we're doing all the time.*

The extent to which Zukov sets off with an agenda to interpret everything in a particular way is obvious from early on in his book, when he comments that in Chinese (*sic*) Wu Li is used to represent physics. Wu can mean matter or energy, and Li can mean universal order, universal law, or organic patterns. Zukov therefore interprets physics as "patterns of organic energy," a description that is beautifully replete with New Age feeling if practically meaningless. What he doesn't do is point out that from his definitions, the term for physics could equally be interpreted as "universal law of energy/matter," which is much more the way it is normally regarded by everyone in the scientific world.

Zukov got very excited about Bell's theorem. At the time he was writing, the first experiments to demonstrate entanglement had been carried out, but Alain Aspect was yet to undertake his groundbreaking work. Even so, Zukov seems to have taken a leap too far. He says that, should the experiments prove the validity of quantum theory, Bell's theorem does away with "local causes." From this, he makes an unwarranted leap from a small, very specific discovery to a major assertion about the universe as a whole.

Zukov assumes that the possibility that entangled particles might exhibit Einstein's "spooky action at a distance" makes it possible to assert that *all* local causality is thrown out of the window. This is despite the fact that, as far as was known at the time, entanglement was a very difficult to construct, fragile state. Everything, he deduces, is connected, all is one—we're back to the Eastern, holistic, knit-your-own-karma ideal of holism. Unfortunately, it *is* a leap too far. Bell's theorem and the experiments to demonstrate its outcome make no suggestion that everything is entangled as Zukov seems to suggest.

Zukov's book was very successful but, though it did bring aspects of modern physics to readers who had never heard of them before, it would also inevitably have made scientists even more wary of quantum entanglement being picked up by the lunatic fringe, which could help explain the response to the Josephson article.

Josephson's fate, however, has not stopped other serious scientists from coming up with dramatic speculations about entanglement, some perhaps even more remarkable, which have not produced anything like the same response.

Of all the entanglement speculations, the most dramatic ideas come from physicist Vlatko Vedral, based until recently at the Blackett

Laboratory of Imperial College, in London. In an article in 2003 in the respected journal *Nature,* Vedral used some new research to launch a remarkable suggestion of a link between entanglement and life itself.

The experiment was conducted by Singhamitra Ghosh, of the James Franck Institute and Department of Physics, University of Chicago, and colleagues there, in London, and in Wisconsin. The surprise result occurred when looking at how the magnetic properties of a lithium salt varied with temperature. The salt was more strongly magnetic than expected—more of the atoms, which act as tiny magnets, were lined up than could be explained using classical physics. But if the calculations were undertaken assuming that the atoms were entangled, the outcome exactly matches the observed results.

It seems that quantum entanglement can influence not only the tiny individual behaviors of pairs of particles, but that of a whole magnetic structure. Here was entanglement making a change to the magnetic strength of something that could be touched and picked up, not an incredibly tiny particle. What's more, Ghosh and his coworkers also showed that other properties of the salt, including its heat capacity, were influenced by entanglement.

As Vedral points out, this shows that entanglement is of paramount importance to the behavior of this substance, behavior on the level that we normally experience and measure; and even a small amount of entanglement can produce significant effects in the human-scale world, the "real" world of tangible physical objects. With this observation in place, Vedral is ready to make a speculative leap.

Quantum theory, he points out, is our most accurate description of how atoms work, and underlies all that we know about chemistry. Chemistry, in turn, provides the operational mechanisms for biology, including those that drive our metabolisms and our reproduction. "So, might it be not only that quantum effects are responsible for the behaviour of inanimate matter, but that the magic of entanglement is also crucial in the existence of life?" Vedral concludes. Our very existence could be made possible by entanglement.

Note this is quite different from Zukov's idea, which took a leap from an experiment involving two particles all the way to "the whole world is like that"—Vedral takes a much more logical, stepwise progress from experimental observation to speculation. Even so, his idea might seem very unlikely, if only because entanglement is difficult to achieve. The entanglement experiments we have seen in previous chapters were initially quite delicate, and the limit, for example, for entangled photons to survive in a fiber optic is only around 20 kilometers (12½ miles), giving them an entangled lifetime of a tiny fraction of a second. If entanglement is so fragile and subject to decoherence, how could it be responsible for as robust and long lasting a phenomenon of life?

The sometimes controversial, but respected British physicist, Sir Roger Penrose, has pointed out that in fact this is almost seeing things back to front. From the nature of quantum particles, entanglement is a normal state of affairs. It's not so much a surprise that entanglement exists, as that it isn't a dominant effect in the physical world. Undisturbed, entanglement should ripple naturally through the quantum particles making up all matter, as a fictional Ice Nine crystal would explosively spread through water.

Ice Nine was a neat concept that appears in the novel *Cat's Cradle* by the highly imaginative American science fiction writer Kurt Vonnegut. He described a newly discovered form of ice, so stable that it only melted at 114 degrees Fahrenheit (45 degrees Celsius). If water ever got into an Ice Nine form, the chances are that under normal weather conditions it would never get out of that form. Should a seed crystal of Ice Nine be dropped into a lake or ocean, it would spread uncontrollably from shore to shore, locking up the water supply and devastating the earth.

Luckily, Ice Nine doesn't exist (though it was a wonderful inspiration), although there is a type of ice that forms at very low temperatures with the intentionally similar name of Ice IX. This, however, isn't stable at room temperatures, and anyway doesn't have the same properties. But this idea of dropping in a seed that rapidly spreads across everything would seem to be the natural state of affairs for entanglement.

The reason entanglement doesn't spread uncontrollably, according to Penrose, is that the act of observation destroys it. We see this in all the applications of entanglement, and Penrose envisages this happening on a much larger scale in the universe. He suggests that nature is continually making measurements on quantum objects, collapsing the entanglement. Roger Penrose isn't suggesting by this that cows are out in the fields with tape measures, or that stars are busy with interferometers, but rather that the impact of all the different objects in the universe on each other—their touching and moving and responding to light—is the equivalent of measurement, and this collapses the otherwise naturally endemic state of entanglement. Other physicists, like Professor Tony Sudbery of York University, suggest it's not so much that observation destroys the entanglement but rather, because we ourselves are part of the entangled world, we can't see beyond our own component of the entanglement.

The most recent of the amazing possibilities for entanglement's role in the universe was also dreamed up by Vlatko Vedral, now the Centenary Professor of Quantum Information Science at the University of Leeds. In 2004, Vedral used entanglement to explain a strange effect in the supercold world, but his explanation also offered a possibility of a fundamental role for entanglement in the nature of matter itself.

When materials are extremely cold, near to the absolute minimum limit of –273.16 degrees Celsius (about –460 degrees Fahrenheit), the point at which atomic motion should cease and temperature no longer strictly exists, their behavior is much more strongly influenced by quantum mechanical effects than at room temperature, leading to seemingly unnatural responses. Some materials, like helium, become superfluids with no viscosity at all. According to Phil Meeson, a physicist at the University of Bristol, "If you were to flap your hand around in a superfluid, it would be like flapping your hand in a vacuum. It's like there is nothing there." Of course, in practice if your hands were near that temperature they would no doubt shatter, but the idea that there is simply no resistance at all to physical motion is the important point.

If you start a ring of superfluid spinning, it will go on spinning forever—there is no friction to stop it. Most famously, superfluids will spookily attempt to climb out of containers as there is no friction to resist the random motion of the molecules. At these incredibly low temperatures, some materials become superconductors that have no electrical resistance, perfectly transmitting an electrical current. And one of the odder behaviors that occur is the Meissner effect, discovered in 1933 by Walter Meissner and Robert Ochsenfeld.

Place a lightweight magnet above a superconductor, and the magnet will levitate, floating in space. The magnet produces an electrical current in the superconductor, which generates its own magnetic field. A simplistic view is that the superconductor is a perfect conductor, so allows the electrical current produced by the magnet to flow perfectly, producing an electromagnet that repels the original magnet. Actually, things are a little more complex, because a moving or changing magnetic field is required to generate a current. Just having a magnet sitting on top of a perfect conductor would not produce the Meissner effect.

What's needed is a magnet sitting on top of a material yet to become a superconductor. The magnet's field penetrates the material. But as the material cools through the critical temperature at which it becomes a superconductor, currents begin to flow on the surface of the new superconductor, effectively pushing the field out the material and taking the magnet with it. The magnet floats. Vedral suggests that this effect is a result of entangled electrons in the surface of the superconductor's giving the photons of the electromagnetic field an effective mass, leaving them struggling through the material as if it were quantum molasses.

If that's true, Vedral has reasoned, perhaps entanglement could also explain the mass of everyday things. The forces that link together particles of matter are usually ascribed to intermediate massless particles, such as the photon. Photons don't just provide us with light and other electromagnetic rays, but are constantly, invisibly skipping between particles of matter.

But if all natural forces were transmitted by massless mediators like photons, they should extend indefinitely—yet in practice, most forces have limited range and most of the mediating particles have mass. This oddity is explained by a mysterious, elusive particle called the Higgs boson.

A bit of background is necessary here. A boson, like a photon, is one of the two types of quantum particle (the other being a fermion such as an electron). Bosons can share quantum states, whereas fermions can't. The Higgs boson is a hypothetical particle that no one has yet seen. There have been plenty of attempts, and occasionally the whole idea has been declared defunct, but as yet proof has not gone either way.

The idea was dreamed up by Peter Higgs, of Edinburgh University, in the early 1960s to explain where mass comes from, and why different particles have such varying mass. Each natural force has a corresponding field, a sort of texture of the environment in which it works, which is communicated by a boson. The electromagnetic field's carrier is the photon. Higgs imagined that there was also a field, the Higgs field, that was responsible for mass. According to this theory, a particle's mass comes from its interactions with a Higgs equivalent of a photon, the Higgs boson.

These Higgs bosons are still controversial. They work very well to explain the current theories and observations, but no one has ever seen one of these so-called "God particles" (they were given this name because of their fundamental role of giving the other particles mass). Back in 2001, there was some speculation that they didn't exist because experiments that should have been revealing them weren't delivering, but there is still widespread acceptance that they may be real—the experimenters say we just haven't got powerful enough particle colliders to capture the elusive Higgs. And governments regularly cancel collider projects because they are so hugely expensive, delaying any possible discovery.

Now, the interesting thing here is that Higgs bosons are supposed to work on mediator particles by excluding them, effectively pushing them out of the way, just as the electrons in the superconductor appear to exclude the photons of the magnetic field—but there is no explanation for why this is happening. Vedral suggests that if Higgs bosons were entangled, this would explain their behavior just as it does the exclusion in the Meissner effect.

This is not a detailed theory at the moment, just a speculation—but, unlike speculation on psi effects, the Higgs boson (itself never yet seen) is regarded as mainstream science, and so this aspect of entanglement, that would mean it is an essential part of the mechanism of objects having mass, is considered an acceptable one—and a fascinating possibility should it prove right. If the Higgs boson, the God particle, is truly "entanglement powered," then combined with entanglement's limitless reach and remarkable consequences, it doesn't seem excessive to call entanglement the God Effect.

Despite seventy years or more of attacks from doubters, entanglement is not going to go away. Every experiment takes us a step closer to realizing just how strange the world is at the quantum level. The natural inclination of even as original a thinker as Albert Einstein was to deny the viability of quantum theory, but this has not proved an acceptable strategy. Quantum theory works. It delivers.

This shouldn't be anything new. The chances are high that, as you read this, you are sitting within a yard of a device making explicit use of quantum technology in its design and construction. It started with optics, then electromagnetics in the development of radar (and eventually the microwave oven). Now, inside your personal computer, cell phone, TV, and more, you are likely to own billions of examples of an explicit practical quantum device—the transistor. Most homes also contain yet another quantum technology—the lasers inside CD and DVD players and recorders, and laser printers. None of these would be possible without the strange phenomena of the quantum world.

Quantum entanglement has the potential to rival the basic quantum effects at the heart of the transistor and the laser, in its influence on the world. Alain Aspect, the scientist who first definitively proved entanglement's defiance of local reality, is unusually forthcoming on this point, in an almost biblically reverent introduction to John Bell's papers on entanglement:

> *I think it is not an exaggeration to say that the realization of the importance of entanglement and the clarification of the quantum description of single objects have been at the root of a* second

> quantum revolution, *and that John Bell was its prophet. And it may well be that this once purely intellectual pursuit will lead to a* new technological revolution.

This is no science fiction writer, but one of the most respected scientists in his field. If Aspect seems to be overstating things, he reminds us, "Who would have imagined the ubiquitous presence of integrated circuits when the first transistor was invented?" Quantum encryption, quantum computing, and quantum teleportation could be just the start of the new quantum revolution. Through quantum entanglement, new quantum effects are exploding onto the practical world scene. There can be little doubt that quantum entanglement, the God Effect, *is* the next big thing.

NOTES

PREFACE

Page ix—Huxley's reference to science as common sense is taken from T. H. Huxley, *Collected Essays,* vol. 4—*The Method of Zadig* (London: Greenwood Press, 1970).

Page ix—Feynman's assertion of the delightful absurdity of nature is taken from Richard Feynman, *QED: The Strange Theory of Light and Matter* (London: Penguin, 1990).

CHAPTER 1: ENTANGLEMENT BEGINS

Page 3—Einstein's "spooky actions at a distance" quote is taken from Max Born, *The Born-Einstein Letters* (London: Macmillan, 1971).

Page 3—Schrödinger's paper with the first use of "entangled" was "Discussion of Probability Relations between Separated Systems," *Proceedings of the Cambridge Philosophical Society* 31 (1935): 555–63 [2].

Page 3—A discussion on the origin of the term "entanglement" and the differences between entanglement and *V* can be found on the Centre for Quantum Computation Web site at http://cam.qubit.org/ under "Quest for the True Origin of Entanglement."

Page 5—Babies' response to action at a distance is described in Roberto Casati, *The Shadow Club* (London: Little, Brown, 2004).

Page 7—Newton on gravity from *Principia Mathematica* [The Mathematical Principles of Natural Philosophy], Book III, General Scholium—Isaac Newton, Andrew Motte, trans. (London: H. D. Symonds, 1803).

Page 7—Modern translation of the Latin *fingo* from *The Principia,* General Scholium Isaac Newton, I. Bernard Cohen and Anne Whitman, trans. (Berkeley: University of California Press, 1999).

Page 9—The suggestion that quantum theory's oddities can be explained by an extra dimension is expressed rather obscurely in T. S. Biro, S. G. Matinyan, and B. Müller, "Chaotic Quantization of Classical Gauge Fields," *Foundations of Physics Letters* 14, no. 5, 471–85.

Page 9—Einstein's criticism of nonlocality is from Max Born, *The Born-Einstein Letters* (London: Macmillan, 1971).

Page 11—Max Planck's comment on quanta quoted in Brian Clegg, *Light Years* (London: Piatkus, 2001).

Page 13—Planck's comment on Einstein to the Prussian Academy quoted in Jeremy Berstein, *Quantum Profiles* (Princeton: Princeton University Press, 1991).

Page 14—Shrek's metaphor of ogres being like onions from the movie *Shrek* (Dreamworks SKG, 2001).

Page 15—Gribbin's criticism of Herbert for calling the planetary model of the atom wrong comes from John Gribbin, *Schrödinger's Kittens* (London: Phoenix, 1996).

Page 18—Momentum is defined as "what [a particle] is doing" in J. C. Polkinghorne, *The Quantum World* (London: Penguin, 1990).

Page 18—The picture of uncertainty as a photograph of a fast moving object was described by Peet Morris in a letter to the author.

Page 20—Anton Zeilinger discussed Einstein's early distaste for randomness in his Fujitsu lecture at Cambridge University (October 2004).

Page 20—Einstein's letter to Born saying he would rather be a cobbler is from Max Born, *The Born-Einstein Letters* (London: Macmillan, 1971).

Page 21—Einstein's letter to Born commenting that God does not play dice from Max Born, *The Born-Einstein Letters* (London: Macmillan, 1971).

Page 22—Franklin attributed the "damned lies and statistics" quote to Disraeli in his autobiography, as cited in the *Oxford Dictionary of Quotations* (Oxford: Oxford University Press, 1979).

Page 25—Niels Bohr, *Collected Works* (J. Kalchar, ed.) volume 6 (Amsterdam: North-Holland, 1985).

Page 29—Pais's account of Bohr's "incantation" summoning Einstein is related in Jeremy Berstein, *Quantum Profiles* (Princeton: Princeton University Press, 1991).

Page 31—The "EPR" paper is A. Einstein, B. Podolsky, and N. Rosen, "Can Quantum Mechanical Description of Physical Reality Be Considered Complete," *Physical Review* 47 (May 15, 1935).

CHAPTER 2: QUANTUM ARMAGEDDON

Page 35—EPR's criterion for reality comes from "Can Quantum Mechanical Description of Physical Reality Be Considered Complete"—A. Einstein, B. Podolsky, and N. Rosen, *Physical Review* 47 (May 15, 1935).

Page 39—Abraham Pais's version of Bohr's response to EPR comes from Abraham Pais, *Niels Bohr's Times* (Oxford: Clarendon Press, 1991).

Page 39—Bohr's response to first hearing about the EPR paper was recounted by Léon Rosenfeld in S. Rozental ed., *Niels Bohr: His Life and Work as Seen by Friends and Colleagues* (Amsterdam: North-Holland, 1967).

Page 41—Bohr's written response to the EPR paper is Niels Bohr, "Can Quantum-Mechanical Description of Physical Reality Be Considered Complete?" *Physical Review* 48.

Page 41—Bohr's remark that action at a distance was "completely incomprehensible" is from Niels Bohr, "Space and Time in Nuclear Physics," Mss 14, March 21, 1935 (Manuscript Collection, Archive for the History of Quantum Physics, American Philosophical Society, Philadelphia).

Page 41—Peat's comparison of the Einstein/Bohr split with Cézanne and a realist painter comes from F. David Peat, *Einstein's Moon* (New York: Contemporary Books, 1990).

Page 42—Schrödinger's paper with the first use of "entangled" was "Discussion of Probability Relations between Separated Systems," *Proceedings of the Cambridge Philosophical Society* 31 (1935): 555–63 [2].

Page 43—Einstein's letter to Born confirming his doubts on quantum theory were still as strong in 1944 from Max Born, *The Born-Einstein Letters* (London: Macmillan, 1971).

Page 44—Einstein's comparison of quantum theory with the thoughts of a paranoic comes from a letter to D. Lipkin dated July 5, 1952 (Einstein Archives).

Page 44—Rosen's denial that EPR was a paradox is from P. Lahti and P. Mittelstädt, *Symposium on the Foundations of Modern Physics: 50 Years of the Einstein-Podolsky-Rosen Gedankenexperiment* (Singapore: World Scientific, 1985).

Page 44—Einstein's comment *"ist mir wurst"* on EPR's use of two states is quoted in A. Fine, *The Shaky Game: Einstein, Realism and the Quantum Theory* (Chicago: University of Chicago Press, 1996).

Page 44—The suggestion that "paradox" implies illogicality or absurdity comes from Andrew Whitaker, *Einstein, Bohr and the Quantum Dilemma* (Cambridge: Cambridge University Press, 1996).

Page 46—TV show *Star Trek, The Next Generation* featured a poker game between holographic reproductions of Newton, Einstein, Hawking, and the show's resident android, Data, in "Decent," part 1, first aired on June 21, 1993.

Page 46—Annie Bell's remark about Sunday suits is quoted in Andrew Whitaker, "John Bell and the Most Profound Discovery in Science," *Physics World* (December 1998).

Page 48—Bell's remark that quantum theory was "rotten" was made to Jeremy Bernstein, and recounted in Jeremy Bernstein, *Quantum Profiles* (Princeton: Princeton University Press, 1991).

Page 48—Bell's remark to Herbert about coming across something hard and clear in a region of woolliness is quoted in Nick Herbert, *Quantum Reality: Beyond the New Physics* (New York: Anchor Books, 1985).

Page 48—Bell's preference for Einstein's views over Bohr's is reported in Jeremy Bernstein, *Quantum Profiles* (Princeton: Princeton University Press, 1991).

Page 48—Bell's feeling that Bohr's explanation of EPR was incoherent in asserted in Andrew Whitaker, "John Bell and the Most Profound Discovery in Science," *Physics World* (December 1998).

CHAPTER 3: TWINS OF LIGHT

Page 55—The debate over priority in devising calculus between Newton and Leibniz is described in Brian Clegg, *A Brief History of Infinity* (London: Constable & Robinson, 2003).

Page 55—The court action between Edison and Swann over the electric light is from Brian Clegg, *Light Years* (London: Piatkus, 2001).

Page 56—Richard Feynman's dismissal of John Clauser's idea of testing Bell's equality was in an interview with Amir Aczel, related in Amir Aczel, *Entanglement* (New York: Four Walls Eight Windows, 2002).

Page 62—Bell's letter to Clauser, pleased at having his inequality tested experimentally, is quoted in Amir Aczel, *Entanglement* (New York: Four Walls Eight Windows, 2002).

Page 63—Bell's sad conclusion that "the reasonable thing just doesn't work" is from Jeremy Bernstein, *Quantum Profiles* (Princeton: Princeton University Press, 1991).

Page 64—The problems of leading edge experimenters are described in John Waller, *Leaps in the Dark* (Oxford: Oxford University Press, 2004).

Page 66—Aspect described the difficulties of producing entangled photons in P. C. W. Davies and J. R. Brown, *The Ghost in the Atom* (Cambridge: Cambridge University Press, 1993).

Page 70—Aspect's comment on Einstein's reaction to his experiment is from P. C. W. Davies and J. R. Brown, *The Ghost in the Atom* (Cambridge: Cambridge University Press, 1993).

Page 76—The letter disputing entanglement's spooky action appeared in *New Scientist* 2471 (October 30, 2004).

Page 76—Bell's paper, "Bertlmann's Socks and the Nature of Reality," appeared in *Journal de Physique,* Colloque C2, suppl. au numero 3, Tome 42 (1981): 41–61.

Page 77—Zeilinger's observations on Bertlmann's socks were made in an interview at the University of Cambridge, Department of Applied Mathematics and Theoretical Physics, October 2004.

Page 79—The distance limits of optical fiber and descriptions of the Vienna team's long-distance outdoor links come from Anton Zeilinger's Fujitsu lecture at Cambridge University (October 2004).

Page 80—Information on the quantum effects of reflection from glass is from Richard Feynman, *QED, The Strange Theory of Light and Matter* (London: Penguin, 1990).

Page 82—The entanglement of rubidium clouds using a beam splitter is described in D. N. Matsukevich and A. Kuzmich, "Quantum State Transfer between Matter and Light," *Science* 306 (2004): 663.

Page 83—Alex Kuzmich's observation that a goal of building quantum repeaters is to establish a link between Washington and New York is taken from *New Scientist* 2471 (October 30, 2004).

Page 83—The first transmission of entangled photons across the Danube is described in M. Aspelmeyer, H. R. Böhm, T. Gyatso, et al., "Long-Distance Free-Space Distribution of Quantum Entanglement," *Science* 301 (2003): 621–23.

Page 84—The long-range through-air transmission of entangled particles in China is detailed in Cheng-Zhi Peng et al., "Experimental Free-Space Distribution of Entangled Photon Pairs over 13 km: Towards Satellite-based Global Quantum Communication," *Physical Review Letters* 94 (2005): 150501.

Page 85—Effects of height above ground level on air pressure from Brian Clegg, *The Complete Flier's Handbook* (London: Pan, 2002).

Page 87—Anton Zeilinger's hope to have an entanglement satellite receiver in place by 2010 is quoted in *Nature* 434 (2005): 1066.

Page 87—The suggestion that entanglement effects are "fluffy bunnies" that won't survive outside the laboratory comes from Eugenie Samuel Reich, "Which Way Is Up?" *New Scientist* (October 2, 2004).

Page 89—The experiment using entanglement to synchronize distant clocks is detailed in Alejandra Valencia, Guiliano Scarcelli, and Yanhua Shih, "Distant Clock Synchronization Using Entangled Photon Pairs," *Applied Physics Letters* 85, no. 13 (2004): 2655–57.

CHAPTER FOUR: A TANGLE OF SECRETS

Page 90—The proposal for the military use of entanglement to communicate with submarines is quoted in Jeremy Bernstein, *Quantum Profiles* (Princeton: Princeton University Press, 1991).

Page 92—For more detail on the history of secret messages see Simon Singh, *The Code Book* (London: Fourth Estate, 1999).

Page 97—A full description of the German Enigma machines used in the Second World War and the cracking of the codes at Bletchley Park can be found in Simon Singh, *The Code Book* (London: Fourth Estate, 1999).

Page 99—Ekert's discovery of the possibility of using entanglement for encryption is taken from an interview with the author.

Page 104—Stephen Wiesner's bemused response to his unforgeable banknote idea is quoted in Simon Singh, *The Code Book* (London: Fourth Estate, 1999).

Page 106—Smollin's account of his work with Charles Bennett appears in "The Early Days of Quantum Cryptography," *IBM Journal of Research and Development* 47, no. 1 (2004): 47–52.

Page 111—The commercial device for capturing keystrokes is Key Katcher, described at www.keykatcher.com.

CHAPTER FIVE: THE BLISH EFFECT

Page 117—Blish's discussion of his story is contained in the critical preface to James Blish, *The Quincunx of Time* (London: Arrow, 1976).

Page 120—Melito's change of *tachygraphe*'s name to *télégraphe* is taken from his *Mémoires* (I, 38), quoted in the *Oxford English Dictionary*.

Page 123—The electric telegraph is described as the Victorian Internet in Tom Standage's book of that title (London: Phoenix, 1999).

Page 126—The description of the fictional Dirac transmitter is taken from James Blish, *The Quincunx of Time* (London: Arrow, 1976).

Page 127—Einstein's discussion of relativity's effect on simultaneity is contained in Albert Einstein, *Relativity: The Special and General Theory* (New York: Dover, 2001).

CHAPTER SIX: THE UNREAL MACHINE

Page 150—Babbage's wish that calculations could be carried out by steam is discussed in James Essinger, *Jacquard's Web* (Oxford: Oxford University Press, 2004).

Page 151—The investment of the British Government in the Difference Engine and other information on Jacquard, Babbage, and Hollerith is taken from James Essinger, *Jacquard's Web* (Oxford: Oxford University Press, 2004).

Page 156—Mike Hally explains the complexities of the early years of electronic computing in *Electronic Brains—Stories from the Dawn of the Computer Age* (London: Granta, 2005).

Page 157—The misleading suggestion that computers can do anything is pointed out in David Harel, *Computers Ltd: What Computers Can Really Do* (Oxford: Oxford University Press, 2004).

Page 157—The origins and details of Moore's Law are detailed on Intel's Web site at www.intel.com/technology/mooreslaw/index.htm.

Page 165—The story of the development of the transistor is told in depth in Michael Riordan and Lillian Hoddeson, *Crystal Fire* (New York: W. W. Norton, 1997).

Page 167—David Deutsch's explanation of his self-aware computer's function comes from P. C. W. Davies and J. R. Brown, *The Ghost in the Atom* (Cambridge: Cambridge University Press, 1993).

Page 168—David Deutsch's groundbreaking paper on quantum computers was "Quantum Theory, the Church-Turing Principle and the Universal Quantum Computer," *Proceedings of the Royal Society of London,* Ser. A, 1985.

Page 169—Tim Spiller likens a qubit to color (against a normal bit's black or white) in Hoi-Kwong Lo, Sandu Popescu, and Tim Spiller, *Introduction to Quantum Computation and Information* (Singapore: World Scientific, 1998).

Page 170—The observation that five hundred qubits can represent more states, each requiring a complex number to represent it, than there are atoms in the universe comes from Michael A. Nielsen and Isaac L. Chuang, *Quantum Computation and Quantum Information* (Cambridge: Cambridge University Press, 2000).

Page 172—For more information on Georg Cantor and his work on infinity, see Brian Clegg, *A Brief History of Infinity* (London: Constable & Robinson, 2003).

Page 184—For further details of the RSA public key encryption algorithm see Simon Singh, *The Code Book* (London: Fourth Estate, 1999).

Page 197—Hau's original experiment bringing light down to 17 meters per second is described in L. V. Hau, S. E. Harris, Z. Dutton, and C. H. Behroozi, "Light Speed Reduction to 17 Metres per Second in an Ultracold Atomic Gas," *Nature* 397 (1999): 594.

Page 198—Hau's light stopping experiment is described in C. Liu, Z. Dutton, C. H. Behroozi, and L. V. Hau, "Observation of Coherent Optical Information Storage in an Atomic Medium Using Halted Light Pulses," *Nature* 409 (2001): 490.

Page 198—Lukin's light stopping experiment is described in A. S. Zibrov, M. Bajcsy, and M. D. Lukin, "Stationary Pulses of Light in an Atomic Medium," *Nature* 426 (2003): 638.

Page 199—Hemmer's experiment stopping light in a yttrium crystal is described in *Physics Review Letters* 88 (2002): 023602.

Page 201—The quantum computing experiment using a pair of hydrogen nuclei is detailed in M. S. Anwar, J. A. Jones, D. Blazina, et al., "Implementation of NMR Quantum Computation with Parahydrogen-Derived High-Purity Quantum States," *Physics Review* A 70 (2004): 032324.

Page 201—The self-contained NMR semiconductor experiment is described in G. Yusa et al., "Controlled Multiple Quantum Coherences of Nuclear Spins in a Nanometre-Scale Device," *Nature* 434 (2005): 1001.

Page 201—The quantum dot experiment is described in J. P. Reithmaier et al., "Strong

Coupling in a Single Quantum Dot–Semiconductor Microcavity System," *Nature* 432 (2004): 197.

Page 203—Details of the ENIAC vacuum tube computer come from *Electronic Brains—Stories from the Dawn of the Computer Age* (London: Granta, 2005).

Page 203—Richard Hughes's comparison of quantum computers with valve (vacuum tube) technology is from Julian Brown, *The Quest for the Quantum Computer* (New York: Touchstone, 2000).

CHAPTER SEVEN: MIRROR, MIRROR

Page 206—Wootters and Zurek's original paper on no cloning was "A Single Quantum Cannot Be Cloned," W. K. Wootters and W. H. Zurek, *Nature* 299, no. 5886 (1982): 802–3.

Page 208—The original teleportation paper was C. Bennett et al., "Teleporting an Unknown Quantum State via Dual Classical and Einstein-Podolsky-Rosen Channels," *Physical Review Letters* 70 (1985).

Page 208—Gilles Brassard's comment on the birth of the teleportation concept is taken from Julian Brown, *The Quest for the Quantum Computer* (New York: Touchstone, 2000).

Page 210—Anton Zeilinger's paper describing the first experimental teleportation was published in *Nature* 390 (December 11, 1997): 575–79.

Page 213—The cesium cloud entanglement is described in B. Julsgaard, A. Kozhekin, and E. S. Polzik, "Experimental Long-lived Entanglement of Two Macroscopic Objects," *Nature* 413 (2001): 400.

Page 213—Artur Ekert's assessment of the difficulties of complete teleportation are from an e-mail to the author in July 2005.

Page 213—The next step from the cesium cloud entanglement is described in E. S. Polzik et al., "Entanglement and Quantum Teleportation with Multi-atom Ensembles," *Philosophical Transactions of the Royal Society A: Mathematical, Physics and Engineering Sciences* 361, no. 1808 (2003): 1391–99.

Page 215—Anton Zeilinger's response to the accusation of wanting to drive a truck through an interferometer is taken from an interview at the University of Cambridge, Department of Applied Mathematics and Theoretical Physics, in October 2004.

Page 217—John Gribbin's prediction of teleportation of an electron within forty years comes from John Gribbin, *Schrödinger's Kittens* (London: Phoenix, 1996).

Page 217—Anton Zeilinger's observation that experimentalists should never say "never" was during an interview at the University of Cambridge, Department of Applied Mathematics and Theoretical Physics, in October 2004.

CHAPTER EIGHT: CURIOUSER AND CURIOUSER

Page 223—The quantum theory of consciousness has been discussed in a number of papers, e.g., M. Jibu and K. Yasue, "What Is Mind? Quantum Field Theory of Evanescent Photons in Brain as Quantum Theory of Consciousness," *Informatica* 21 (1997): 471–90.

Page 225—Josephson is quoted on Bell's inequality in P. C. W. Davies and J. R. Brown, *The Ghost in the Atom* (Cambridge: Cambridge University Press, 1993).

Page 225—Information on Royal Mail stamps celebrating the Nobel Prize taken from Royal Mail press release dated August 15, 2001.

Page 226—Text of booklet from Royal Mail Nobel Prize stamps, issued on October 2, 2001. Reproduced with the permission of Royal Mail.

Page 227—News item on Royal Mail Nobel Prize stamps appeared in *Nature* 413 (September 27, 2001): 339.

Page 227—News item on Royal Mail Nobel Prize stamps appeared in *The Observer* (September 30, 2001).

Page 228—James Randi's view of the lack of link between quantum theory and telepathy was broadcast as part of an interview with Brian Josephson on BBC Radio 4's *Today,* October 2, 2001.

Page 232—The "magic square" entanglement game is a simplification of Padmanabhan Aravind, "Bell's Theorem without Inequalities and Only Two Distant Observers," *Foundations of Physics Letters* 15, no. 4 (2002): 397–405.

Page 235—Michael Shermer's criticism of New Age use of quantum labels is in *Scientific American,* January 2005.

Page 236—John Bell's criticism of the idea that physicists spend time discussing mysticism is quoted in Jeremy Bernstein, *Quantum Profiles* (Princeton: Princeton University Press, 1991).

Page 238—Vlatko Vedral's suggestion that life itself might depend on entanglement comes from Vlatko Vedral, "Entanglement Hits the Big Time," *Nature* 425 (2003): 28–29.

Page 239—Roger Penrose's views on the absence of universal entanglement are from Roger Penrose, *The Road to Reality* (London: Jonathan Cape, 2004).

Page 239—The fictional Ice Nine appears in Kurt Vonnegut, *Cat's Cradle* (London: Gollancz, 1963).

Page 240—The actual Ice IX is described in E. Whalley, J. B. R. Heath, and D. W. Davidson, "Ice IX: An Antiferroelectric Phase Related to Ice III," *Journal of Chemical Physics* 48 (1968): 2362–70.

Page 241—Phil Meeson's comment about waving your hands in a superfluid is taken from an interview in *The Guardian* (October 8, 2003).

Page 242—Vlatko Vedral's speculation on entanglement's responsibility for mass is in

Vlatko Vedral, "The Meissner Effect and Massive Particles as Witnesses of Macroscopic Entanglement," www.archive.orgquant-ph/0410021.

Page 244—Alain Aspect's speculation on the technological revolution emerging from quantum entanglement is part of his introduction to John Bell, *Speakable and Unspeakable in Quantum Mechanics* (Cambridge: Cambridge University Press, 2004).

SELECT BIBLIOGRAPHY

Putting together this book involved the great pleasure of reading many books, papers, and journals. It is impossible to list everything that has impacted upon the contents; this bibliography highlights texts that may be of interest to the reader. See the popular science Web site, www.popularscience.co.uk, for more information on the *Popular Science* titles, and for features on new developments in quantum theory.

Aczel, Amir. *Entanglement.* New York: Four Walls Eight Windows, 2002.

Bell, J. S. *Speakable & Unspeakable in Quantum Mechanics.* Cambridge: Cambridge University Press, 2004.

Bernstein, Jeremy. *Quantum Profiles.* Princeton: Princeton University Press, 1991.

Born, Max. *The Born-Einstein Letters.* London: Macmillan, 1971.

Brown, Julian. *The Quest for the Quantum Computer.* New York: Touchstone, 2000.

Clegg, Brian. *Light Years.* London: Piatkus, 2001.

Davies, P. C. W., and J. R. Brown. *The Ghost in the Atom.* Cambridge: Cambridge University Press, 1993.

Essinger, James. *Jacquard's Web.* Oxford: Oxford University Press, 2004.

Feynman, Richard. *QED: The Strange Theory of Light and Matter.* London: Penguin, 1990.

Gleick, James. *Isaac Newton.* New York: Pantheon Books, 2003.

Gribbin, John. *Schrödinger's Kittens.* London: Phoenix, 1996.

Harel, David. *Computers Ltd: What Computers Really Can't Do.* Oxford: Oxford University Press, 2004.

Newton, Isaac. *The Principia.* Berkeley: University of California Press, 1999.

Peat, F. David. *Einstein's Moon.* New York: Contemporary Books, 1990.

Polkinghorne, J. C. *The Quantum World.* London: Penguin, 1990.

Standage, Tom. *The Victorian Internet.* London: Phoenix, 1999.

Whitaker, Andrew. *Einstein, Bohr and the Quantum Dilemma.* Cambridge: Cambridge University Press, 1996.

White, Michael. *Isaac Newton: The Last Sorcerer*. London: Fourth Estate, 1997.

White, Michael, and John Gribbin. *Einstein—A Life in Science.* London: Simon & Schuster, 1994.

Wolpert, Lewis. *The Unnatural Nature of Science.* London: Faber, 1993.

ACKNOWLEDGMENTS

My thanks have to be expressed to the many people who have made this book possible. In practical terms, this has to start with my agent, Peter Cox, and my editor, Ethan Friedman.

Then there are the many people who have provided information, answered dumb questions, and pointed me in the right direction. Thanks in particular to Professor Tony Sudbery, Doctor Peet Morris, Professor Andrew Whitaker, Professor Artur Ekert, Professor Brian Josephson, Professor Anton Zeilinger, Doctor Lov Grover, Doctor Helmut Jakubowicz, Doctor Marcus Chown, Professor Günter Nimtz, Professor David Hough, Ken Bates, Clare Sudbery, Doctor Jason Hinson, Howard High, and the staff of the Cambridge Centre for Quantum Computation.

Finally, I have to say thanks to Gillian, Rebecca, and Chelsea for putting up with the usual irritations and entanglements associated with having a writer about the house.

INDEX